Game Theory and Public Policy,
Second Edition

Game Theory and Public Policy, Second Edition

Roger A. McCain

Professor, School of Economics, LeBow College of Business, Drexel University, USA

Cheltenham, UK • Northampton, MA, USA

Published by
Edward Elgar Publishing Limited
The Lypiatts
15 Lansdown Road
Cheltenham
Glos GL50 2JA
UK

Edward Elgar Publishing, Inc.
William Pratt House
9 Dewey Court
Northampton
Massachusetts 01060
USA

Paperback edition 2017

A catalogue record for this book
is available from the British Library

Library of Congress Control Number: 2015945467

This book is available electronically in the **Elgar**online
Economics subject collection
DOI 10.4337/9781784710903

ISBN 978 1 78471 089 7 (cased)
ISBN 978 1 78471 090 3 (eBook)
ISBN 978 1 78811 212 3 (paperback)

Typeset by Servis Filmsetting Ltd, Stockport, Cheshire
Printed and bound in Great Britain by TJ International Ltd, Padstow

Contents

Preface to the revised edition

The first edition of *Game Theory and Public Policy* arose from the exploration of some simple examples that attempted to represent some issues of public policy, including externalities, in terms of noncooperative and cooperative game theory. Thus, the first edition comprised two parts: the first a survey of models from the received literature of game theory with a view to their application to public policy, and the second an exposition of some proposed refinements of models used to represent externalities in particular. However, the material of Part II has been superseded by my subsequent research, largely published in McCain (2013). Thus, Part II has been completely rewritten. This in turn suggested some revisions of Part I, principally shifts of emphasis from topics that had served mainly to introduce ideas that were expanded in Part II. In addition, there have of course been some updates to reflect other developments in game theory and economics since 2008.

PART I

Historical and critical survey

1. Objectives and scope of the book

In recent years game theory has become more prominent as an aspect of research and applications in public policy disciplines such as economics, philosophy, management, and political science, and in work within public policy itself. One reason for this growing prominence may be understood from some comments of Thomas Schelling (1960) and Robert Aumann (for example, 2004). They have said that the subject matter of game theory would be better described as interactive decision theory. Schelling and Aumann shared the Nobel Memorial prize in 2005 for their work in game theory (Royal Swedish Academy of Sciences), and Aumann was the first president of the world Game Theory Society.

Why then use the term "game theory" for a field that is not really about games? The *game* is to *game theory* as the *experiment* is to *experimental science*. After all, experimental science is not about experiments. It is about the natural world. Nevertheless experiments are a powerful aid to our understanding of the natural world. Similarly, when we conceive interactive decisions as games, we have a powerful aid to understanding them (and among other things, to the design of experiments).

Game theory is, as Aumann says, an interdisciplinary field. "There are very few subjects that have such a broad, interdisciplinary sweep. Let me just put over here some of the ordinary disciplines that are involved in game theory. We have mathematics, computer science, economics, biology, (national) political science, international relations, social psychology, management, business, accounting, law, philosophy, statistics. Even literary criticism . . . We have sports" (Aumann, 2003, p.4). Of course, none of these disciplines depends on game theory for its existence. Nevertheless, game theory can be set apart as an attempt to understand collective human activity as the outcome of interactive decisions. On the one hand, this is a remarkably ambitious venture. On the other hand, to the extent that it is successful, it must surely be a crucial foundation for the study of public policy.

The objective of this book is to survey and advance our understanding of game theory as a tool of public policy analysis. The hope is to advance that understanding less by the statement and proof of broad theorems (although the value of such proofs is not to be minimized, and will play some role) as by the clarification and critical assessment of the theorems we have, and by multiplication of examples and survey and extension of specific cases of application. In practice, the influence of game theory on public policy and the related disciplines has been less a consequence of broad theorems than of insightful examples. Accordingly, it is hoped that a critical reconsideration of some of those examples, and discussion of some new or less-known ones, will contribute to the study and ultimately the practice of public policy.

Public policy is a pragmatic field. The pragmatic perspective leads to a view of public policy as an outcome of a *process*, and public policy *analysis* is often carried on in terms of the *public policy process*. We might sketch the public policy process roughly as follows: (1) A *problem* is identified which seems to call for public initiative as a solution. (2) Alternative solutions are proposed. (3) Solutions are evaluated, and to the extent possible, the most promising solution specified. At this point the process may be abandoned, if it is found that the best solution does not require public initiatives. We should note, too, that different individuals with different values or interests may regard different proposals as best, and this is the stuff of which politics is made. From this point we suppose that one particular political perspective has been adopted, and the proposal is considered best from that particular perspective. (4) The proposal is advocated and public support for it sought, in the course of which new interest groups and organizations may come into being. (5) The proposal is brought before the legislative or executive branch of government at an appropriate level. (6) The proposal is enacted with or without modification. (7) The proposal is implemented. (8) Experience with the program as implemented leads to feedback from those affected. (9) The cycle begins again with proposals for improvement, replacement, or abandonment of the policy.

How will game theory fit into this outline? It is widely understood today that there are two great branches of game theory, the noncooperative and the cooperative branch. Of the two, noncooperative game theory has been the more influential, especially in the last quarter of the twentieth century and the beginning of the twenty-first. This is often treated as an institutional dichotomy: cooperative game theory is applicable when agreements are enforceable, while noncooperative game theory is applicable otherwise. This book will argue that, on the contrary, the two branches of game theory reflect different conceptions of rationality. Moreover, neither conception is altogether satisfactory. The book argues that noncooperative game theory is effective as a problem-finding or diagnostic method – noncooperative behavior is common enough so that a social arrangement that is unstable in the face of noncooperative behavior will probably fail. However, solutions based on noncooperative game theory may be unstable in the face of cooperative or collusive behavior, and cooperative behavior is common enough that such solutions will themselves often fail.

This should be qualified in the following way, however. There is also some research that combines cooperative and noncooperative game theory, and one particular branch is sometimes called social mechanism design.[1] If game theory is interactive decision theory, we may think of the outcome of the interaction as being jointly determined by the decisions and the "rules of the game." In social mechanism design, a particular goal for action is specified, and the objective is to find "rules of the game" that will make the goal the outcome of the game assuming noncooperative or cooperative decisions as the case may be. In the context of social mechanism design, noncooperative game theory may also be useful at the second and third stages, proposal and evaluation of new policies. There have been some successes in this way, but also some failures, with both occurring in particular in the design of public auctions of electromagnetic spectrum for telecommunications.

In game theory a state of the game that meets certain conditions, such as stability in some specific sense, may be a candidate *solution* of the game. The word solution is meant in a mathematical rather than a pragmatic sense, here. An array of decisions that is stable in the sense that no-one can improve his outcome by changing his strategy unilaterally (while others continue their strategy decisions unchanged) is a Nash equilibrium, and the Nash equilibrium is probably the best known and most widely applied concept of noncooperative solution.

In cooperative game theory, binding agreements to choose a common decision or a joint strategy are considered to be possible. A group that makes such an agreement is said to form a "coalition." The word "coalition" is best known in its political usage, as a group of political parties in a parliamentary government who join together to form a majority and govern jointly. In cooperative game theory, the word has been generalized to refer to any group of players in a "game" who join together to choose their strategies jointly. Most games with more than two players, applicable to problems of public policy, will provide cases in which individual actors could benefit by forming coalitions with binding agreements to choose a joint strategy. Indeed, as Maskin (2004) points out,[2] "we live our lives in coalitions." Thus an account of social life (and especially of public policy) that ignores cooperative game theory must be quite incomplete.

In any case, the formation of coalitions will be crucial at stages 4–6 of the public policy process as sketched above. Coalitions are likely to be important at other stages as well. Noncooperative game theory can fail because it assumes that people act noncooperatively when in fact they can and do form coalitions, such as bidding cartels in auctions. Therefore cooperative game theory may be essential at stages 2 and 3 as well. We acknowledged that stage 3, in particular, would be dependent on values and interests that might differ. Even when that is so, there may be scope for the differences to be accommodated and the distinct interests and values to be advanced jointly. That, too, is the stuff that politics is made of, and it is also a subject of cooperative game theory. We cannot avoid the conclusion that cooperative game theory is essential for a complete understanding of public policy.

This presents a number of difficulties. First, there are several concepts of solution in cooperative game theory. Which (if any) will be most helpful for our purposes? Second, much of the literature relies on powerful simplifying assumptions. Such simplifying assumptions permit the statement and proof of broad and powerful mathematical theorems, but at the same time they indicate the limits of the applicability of the theorems. Together, these simplifying assumptions mean that most cooperative game theory is not applicable to very many problems of public policy. To be specific,

(1) Expressing the game in the simplified coalition function form means that it cannot be applied to any case in which there are *externalities* and consequent *inefficiencies.*

(2) The common assumption of superadditivity means that if agents are rational, the grand coalition will always form and will efficiently determine the strategies of every agent. This means it is simply not applicable to any case in which decentralization is persistent.

(3) The world we observe, the world relevant to public policy, seems to be one in which many coalitions form and often act independently and indeed competitively with one another. In cooperative game theory such an array of distinct coalitions is called a "coalition structure" (Aumann and Dreze, 1974). We would like a theory that would give us some insight as to just what coalition structures would be likely to form, and why, and game theory based on coalition functions and superadditivity is not helpful with that.

There are approaches to cooperative game theory (as we will discuss in the next chapter) that allow both for externalities and coalition structures, but these approaches are "mathematically intractable." That is, they probably do not have very general solutions, and if they do, the solutions are very hard to find and only tentative progress has been made in this direction. We might nevertheless find solutions for particular cases, and even develop a tool-kit for seeking such special-case solutions.

The objective of the book, then, will be a critical review of some major topics from both cooperative and noncooperative game theory, including some less known ideas in noncooperative game theory, and some constructive proposals for new approaches, to assemble a tool-kit for the analysis of public policy, with the pragmatic purpose of identifying problems and exploring potential solutions. At the same time, we may find resources for a clearer understanding of the public policy enterprise itself.

NOTES

1. The 2007 Nobel Memorial Prize honored contributions of this sort, as did the 2012 Nobel.
2. This is a quotation from memory from Maskin's address.

2. Representing games

The first step in any application of game theory, whether to public policy or for any other purpose, is to represent the real-world phenomenon of interest as a problem of interactive decision, that is, a "game." This chapter will set out some forms for representation of games that will be important for the remainder of the book. Some will be familiar, even pedestrian, to the reader who is well grounded in game theory. Nevertheless some topics may be important for the game theorist, if only for differences of stress. Contingent strategies are well known, but this book will often make them more explicit and formal than they often are in the game theory literature. Nested games may be a novel topic to the game theorist, as the concept comes from applications in political science, and are crucial to the distinction of a private from a public sector. "Imperfect recall" is very little mentioned in recent game theory, and needs to be discussed in the context of cooperative game theory. Finally, the discussion of externalities in cooperative game theory may be novel to some game theorists. These are important concepts for public policy. Nevertheless, the chapter is expository, with nothing new to the literature except specific examples, some terminology, emphasis and expression.

2.1 GENERAL CONSIDERATIONS

Game theory is a (mathematically) formal study, with deep roots in mathematical set theory. The language of set theory is designed for generality even at the expense of intuition and common sense. For example, in set theory we routinely speak of a set without any members, the "null set," or a set with only one member, or a set consisting of all of the members of some population. These usages may seem strange from the point of view of ordinary English. Generally words come to us with connotations as well as formal definitions, and a word like "set" tends to connote a plural grouping within some larger grouping. Thus the idea of a set without members may seem silly, and the natural impulse of the reader is to make the charitable assumption that the author is not silly, so that something else must be meant. In the present context, this charitable impulse is likely to cause confusion. Assume instead that I am silly. Because of these conventions, though, some of the most important and productive forms of representation are distant from intuition.

Since a game is an interactive decision problem, our representation must include at least a set of decision makers, some alternatives among which they must decide, and some objectives to be advanced by the decision. Let us call the set of decision makers N and denote the members of the set of decision makers $i = 1, 2, \ldots n$. The set of decision makers is non-empty; that is, it has at least one member. We will usually refer to the members of this set as "players" or "agents." We will sometimes talk about "games" with only one player, although in that case there is no interaction. The objectives for different players will usually be different, and may be conflicting. For now, we will simply represent those objectives as numbers, and think of the numbers as money payoffs from the "game."

Here is an example of interactive decision theory. (We will call it Game 2.1, the Water Game.) Eastland and Westria share the valley of Southflowing River, which forms the boundary between them. Each country controls some of the northern tributaries of the river, and could divert water from the tributary streams for their own use. However, any diversion from the tributaries of the river will divert water that the citizens of the southern regions of both countries use for irrigation and other purposes, and if both countries divert the water of the tributaries, the flow in the south will be so reduced that silting and problems of navigation will also occur. Reliable cost–benefit studies have provided the following figures: if just one country diverts water from the tributaries, the net benefit to that country will be 3 billion euros, but the other country will lose 4 billion. However, if both countries divert water from the tributaries, each country will suffer a net loss of 2 billion. The two countries do not trust one another and keep their decisions strictly secret from one another as long as possible, so each country can only conjecture as to what the other country will decide and feels no possibility of influencing the decision of the other. In this example, the players are the two countries, and the alternatives are different ways of obtaining water for each country's needs. The decisions are to divert water from the tributaries or not. The cost–benefit studies establish that the decisions are interactive: that is, each country's net benefits or losses depend on the other country's decision as well as their own. This example illustrates the simplest class of nontrivial games, two-by-two games; that is, two-player by two-strategy games.

2.2   THE GAME IN EXTENSIVE FORM

The most intuitive way to represent a complicated decision problem is as a tree diagram, in which each decision is represented by a branch in the tree. In our example, the two countries make their decisions more or less simultaneously, each in secrecy. This lack of information should also be represented in the tree diagram. That's a complication we shall leave for a little later. First consider the following, somewhat simpler example: Game 2.2, the Entry Game.

One of the most important of simple games, both for theory and for applications in economics and public policy, is the game of market entry. Firm A is an established monopolist, and Firm B is a firm considering entry into competition with Firm A. Firm B has two choices: it can enter or not. Firm A then has two choices: it can retaliate against the entrant by means of a price war, or it can accommodate the new firm by maintaining a price that will be profitable to both of them. Either way, Firm A will face lower profits, but the price war results in even lower profits than the strategy of accommodation. Game 2.2 illustrates this case with payoffs on a scale of 5, and the first payoff to the entering firm, Firm B, and the second payoff to Firm A. (The reader may add as many zeros as seems realistic.)

Figure 2.1 shows this game in the form of a tree diagram, reading from left (the root) to right (the branches). The first number at the tip of each branch is the payoff to Firm B, and the second to Firm A.

Figure 2.2 represents the Water Game in the tree diagram form conventional in game theory. The first payoff is to Eastland and the second to Westria. We see that Westria's decision is enclosed in a larger lozenge

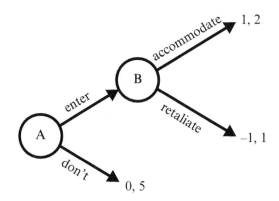

Figure 2.1   Game 2.2: the Entry Game

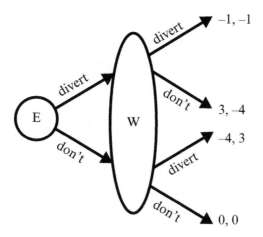

Figure 2.2   Game 2.1: the Water Game in extensive form

that includes both of the branches that come from Eastland's decision. This lozenge is called an "information set,"[1] and it encodes the fact that Westria does not know at which branch it is (which decision Eastland has made) when it makes its decision. Conversely, the node labeled "E" in Figure 2.2 is a *full information node*, as are "A" and "B" in Figure 2.1. At a full information node the player is aware of commitments already made by the other players, if any. For a game like the Water Game, in which the decisions are made simultaneously, either decision maker can be taken first, provided that the information available to each decision maker is accurately represented.

When games are represented as tree diagrams with information sets to indicate decisions made under ignorance, the game is said to be represented *in extensive form*. The extensive form is highly intuitive, but is not the representation most often used in game theory.

2.3   THE GAME IN STRATEGIC NORMAL FORM

Continuing to use the example of the Water Game, we may instead represent it in tabular form. Consider Table 2.1. By choosing to divert or not, Eastland determines whether the outcome will be the payoffs in the first or second of the bottom two rows of the table. Westria's decision whether to divert or not determines whether the outcome will be the payoffs in the last or next to last column. In each cell the first payoff is to Eastland and the second to Westria.[2] Putting all this together, the table tells us (for example)

Table 2.1   Game 2.1 in strategic normal form

Payoff order: Eastland, Westria		Westria	
		Divert	Don't
Eastland	Divert	$-1,-1$	$3,-4$
	Don't	$-4,3$	$0,0$

that if Eastland diverts and Westria does not, the net benefits are as shown in the upper right cell, a loss of 4 for Westria and a net gain of 3 for Eastland. This presentation will be familiar to those acquainted with the Prisoner's Dilemma example. When the game is presented in a tabular form such as this, the game is said to be represented in *strategic normal form* or, more briefly, in *normal form* or in *strategic form*. Fairly obvious extensions of the tabular presentation can be used with games in which there are more than two strategies or in which there are three or even four players.

For larger games, a more mathematical presentation is necessary, and in general we may borrow language from set theory. Letting S_i be the set of all strategies available to player i of n in the game, Σ be a set with n elements of which element i, which we may call σ_i, is an element of S_i (that is, σ_i is the strategy chosen from S_i by agent i), and $\mathbf{v} = (v_1, v_2, \ldots, v_i, \ldots v_N)$ be a vector of the n payoffs to the n players in the game. In place of the table we have a function that gives a vector of payoffs $\mathbf{v} = f(\Sigma)$, with one payoff for each of the players i, corresponding to each possible set of strategies Σ. Then the game (in strategic normal form) is said to comprise the set of players N, the set of strategies S_i for each player, and the payoff function f. In general, any table is just a visual way of presenting a mathematical correspondence, and that is true equally of the payoff table in game theory as of other tables.

Von Neumann and Morgenstern (1994/2004) proved that any game in extensive form can be represented also in strategic normal form, and in particular, the Entry Game can be represented in that way. However, there is a trick to it, and the trick is often neglected, even in game theory research that is quite advanced in some ways. For Firm B, the case is similar to the case for Eastland and Westria: Firm B simply has to choose between two actions, enter or don't. Firm A, however, knows Firm B's decision when it makes its own decision, and Firm A's decision is conditional on that knowledge. This is a *contingent strategy*. For firm A there are four contingent strategies:

Strategy 1: "If Firm B enters, then retaliate, otherwise retaliate."
Strategy 2: "If Firm B enters, then accommodate, otherwise retaliate."

Table 2.2   Game 2.2 revised: the Entry Game in strategic normal form

Payoff Order: Firm B, A		Firm A			
		1	2	3	4
B	Enter	−1,1	1,2	1,2	−1,1
	Don't	0,5	0,5	0,5	0,5

Strategy 3: "If   Firm   B   enters,   then   accommodate,   otherwise accommodate."

Strategy 4: "If Firm B enters, then retaliate, otherwise accommodate."

This list of contingent strategies may seem trivial, redundant, and even downright silly, by comparison with a simple enumeration of the choices to retaliate or accommodate, but all are contingent strategies that are available to Firm A in the light of the information it has when it makes the decision. Therefore, all are necessary for a complete presentation of the game in strategic normal form as defined by von Neumann and Morgenstern. Table 2.2 shows Game 2.2 in strategic normal form.

It is common to refer to decision alternatives such as "enter" and "don't enter" and "accommodate" or "retaliate" as "strategies," but this is not consistent with the representation of the game in strategic normal form as understood by von Neumann and Morgenstern. The extensive form was clarified in a key early paper of Kuhn (1997, pp. 46–68). In it Kuhn distinguished between strategies as conceived by von Neumann and Morgenstern and what he called "behavior strategies," that is, local decisions in the different decision nodes of the tree. It has become common not to make that distinction, and in some cases confusion can result. For this book, the decision alternatives such as "enter" and "don't enter" and "accommodate" or "retaliate" will be called *behavior strategies*[3] and strategies such as strategies 1, 2, 3, and 4 above will be called *contingent strategies*. I will make every effort not to use the term "strategy," without modification, unless the meaning will be clear from the context.

Here is an example that will illustrate the importance of distinguishing contingent from behavioral strategies. To give the example a real-world background, consider a case in allocation of intellectual property. Firms A and C have patents on alternative methods of producing widgets. Firm C is an established monopolist but the cost of production with their patented technology is relatively high. The technologies are complementary, so that a company in possession of both technologies could be a low-cost producer in the widget market. Firm B is known to be interested in entering the

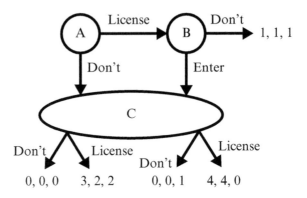

Figure 2.3   Game 2.3: Selten's "Horse"

widget market and has applied to license Firm C's patent. Firm C does not know whether or not Firm A has licensed its patent to Firm B, but Firm B's application to Firm C for a license would make sense only if Firm B intends to enter. If B holds the license for A's patent, and does not enter the widget market, then each firm can continue in its traditional markets, and we represent this outcome by payoffs of 1,1,1 (in the order of Firm A, B, C). If Firm B does enter, and does so with a license only for Firm A's technology, the market will be shared profitably for all firms; we represent this by payoffs 3,3,2. If Firm B enters with neither license, its competition will leave all three firms without profits, which we represent by payoffs 0,0,0. If Firm B enters with only the license for A's patent, then it will be unprofitable and unable to pay license fees to Firm A, so that those firms will be unprofitable, but Firm C will continue with some profits, giving payoffs 0,0,1. Finally, if B enters with both licenses, it will become dominant as the low-cost producer in the widget market, so that Firms A and B will be highly profitable but Firm C unprofitable, for payoffs 4,4,0. Again, the reader may multiply these payoff numbers by a factor large enough to make them "realistic."

Thus we have the game shown in Figure 2.3. With this sequence of decisions and payoff numbers we have replicated an example of Reinhard Selten's[4] (1975). It is called "Selten's Horse," because of a slight resemblance of the diagram for the game in extensive form to a horse. Accordingly, we will follow Selten's notation and denote the behavior strategies as follows:

- Player A. License = R1, don't = L1
- Player B. Enter = L2, Don't = R2
- Player C. Don't = L3, License = R3

Table 2.3   Game 2.3: a normal form representation of Selten's "Horse"

Payoff order Firm A, B, C	Firm A			
	Don't		License	
	Firm B		Firm B	
	If License then Don't else no action	If License then Enter else no action	If License then Don't else no action	If License then Enter else no action
C If Don't or License, Enter then Don't else no action	0,0,0	0,0,0	1,1,1	0,0,1
If Don't or License, Enter then License else no action	3,2,2	3,2,2	1,1,1	4,4,0

This game is presented in a valuable advanced textbook as an example where some standard methods break down (Montet and Serra, 2003). It illustrates a case in which some decisions must be made with very limited information, which is the case for Firm C. The information available to Firms A, B, C, is complex but clear enough from the diagram. Firm B, the second player, knows what Firm A has done, since he would have no opportunity to choose had Firm A not chosen to license its technology. For Firm C, who plays last, there are three possible sequences of choice by the previous two players. They are

1. L1
2. R1, R2
3. R1, L2

Of these, Firm C can rule out only the second – the first and third are equally possibilities for him. In order accurately to represent this game in normal form, we have to preserve this information structure. To do this it is necessary to use the contingent strategies.

The normal form representation of the Horse Game is shown as Table 2.3. In order to represent this three-person game, the strategy choice of Firm A determines which side of the table the other two players play in. If firm A chooses "License," then the other two play in the right side of the table; otherwise, they play in the left. Montet and Serra suggest that this game is particularly problematic for analysis in terms of *behavior strategies*, since

(1) Firm C has no rational basis to choose either of the behavior strategies "Don't" or "License," and therefore (2) Firms A and B, unable to anticipate the choice that will be made by Firm C, also do not have any rational basis for their own choices. However, Selten (1975) found a rational-action solution for the game represented in strategic normal form with contingent strategies. This solution will be discussed in Chapter 6. For further discussion of contingent strategies and further examples see Dutta (1999, Chapter 2) and McCain (2014b, Chapter 2).

When the game is represented in extensive form, we define a *subgame* as a full-information node (such as A in Figure 2.1) together with all possible moves that follow it in the game tree (as "accommodate" and "retaliate" follow A in Figure 2.1). In a larger, more complex game, subgames could themselves be quite large and complex. For generality, it is conventional to regard the whole game (beginning with B in Figure 2.1) as one of its subgames; smaller subgames are called "proper subgames." Notice that the "Horse" Game has no proper subgames. These concepts are standard in introductory texts, where many more examples can be found; see, for example, McCain (2014b, Chapter 12) and Dutta (1999, Chapter 11).

Following Tsebelis (1990), we may denote any sequence of moves that is part of a game in extensive form as a *nested game*. If the sequence of moves is a subgame, then we will denote it as an *imbedded game*. Examples may be found in Chapter 6 below, and see also McCain (2014b, Chapter 13). These concepts will be important for our purposes since the activities of the private sector in a market economy will always be nested in the larger game that includes the determination of public policy. If, in addition, the private sector activities constitute an imbedded game, that is, a subgame of the larger game, then noncooperative game theory can validly be applied to them. If the private sector activities are nested but not imbedded, then we will need to be more careful (as Tsebelis observes).

The strategic normal form representation has been central to most applications of noncooperative game theory, and will be extensively discussed in Chapters 4 and 5.

2.4 UNCERTAINITY AND CALIBRATION

While most noncooperative game theory is based on "games" with numerical payoffs, this needs to be qualified in two ways. First, it is recognized that the actual benefits attached to decisions are subjective benefits, and that the numbers ideally should be quantities of subjective satisfaction, that is, in the language of twentieth-century economics (and earlier utilitarianism) the payoffs are quantities of "utility." Second, payoffs may be uncertain from the point of view of the player, the theorist, or both. In the Water Game, as we have seen, Westria must make their decision in what is called an "information set" but might better be thought of as an ignorance set – Westria does not know what decision Eastland will have made. In this example, as in game theory in general, the interactive decisions have to be made under *uncertainty*. Put otherwise, Westria chooses between two strategies, not on the basis of their payoffs, but on the basis of some probability distribution over the payoffs, which in turn depends on a probability distribution over the strategies of others. The simplest way to deal with this problem is also the one usually used: we assume that the decision maker wants to choose the strategy that yields the largest mathematical expectation of payoffs. Thus probability and mathematical expectations play a central role in game theory. In much of the research literature it is taken for granted that the objects of choice are probability distributions over strategies and payoffs, and the literature is hardly intelligible if this is not kept in mind.

Then, (a frequent question in beginning classes) how do we figure out what the payoff numbers should be? Generically, game theory models are not usually calibrated precisely. In some applications, the application itself may provide evidence that can be used for calibration. The Water Game is an example in which (hypothetically) the calibration is derived from cost–benefit analysis in the specific case. Very commonly the payoffs are treated as algebraic unknowns, with some restrictions on their values. It is fairly common to find that the payoffs can vary over some positive range without affecting the qualitative results, while payoff values outside that range have very different results. In the Water Game, for example, all the payoffs can be multiplied by any positive constant number, or have any constant number added to them, or both, and the rational decisions of the two players in the game will not change. Thus exact calibration is often not necessary, and may not be very helpful, and the numbers chosen for illustrative purposes can be arbitrary, so long as they are within appropriate ranges.

2.5 COOPERATIVE GAMES

The examples we have seen so far are drawn from "noncooperative" game theory. The Water Game, in which the agents act with deliberate secrecy and distrust one another, illustrates a noncooperative game particularly well. Cooperative game theory is applicable whenever the players in a game can form "coalitions," that is, groups that choose a common strategy to improve the payoffs to the members of the group. Cooperative game theory relies especially on mathematical set theory for many of its basic ideas. The usual assumption is that any group of agents in the "game" can form a coalition, and a coalition among agents a, b, and c would be denoted as $\{a, b, c\}$. The brackets $\{\}$ are conventional in set theory to indicate the "elements" of a "set," or in alternative ordinary language, the individuals who make up a grouping.

Most studies in cooperative game theory will begin with an enumeration of all possible coalitions. As before, suppose there are n "players in the game." An individual agent can be indicated by a_i, or simply by the index i, with $i = 1, \ldots n$. Common sense would see that any group of agents with more than one and less than n could form a coalition (with more or less difficulty). That's right, but it is not complete. In addition to those groupings, we also enumerate all *singleton coalitions*, $\{a_i\}$, that is, "coalitions" with just one member, and also the *grand coalition* of all n agents in the game and the null coalition, $\varnothing$, a "coalition" with no members. (By convention in set theory $\varnothing$ means a set with no members.)

When a coalition is formed, the expectation is that by working together and choosing a joint strategy they will be able to improve their results overall. It may be that one member, let us say c, bears a special cost for this, or another agent, such as a, gets most of the benefit. An example might be the modification of a river course, so that those upstream benefit (from the diverted water) but those downstream lose. Then c is downstream and a is upstream. To enlist c in the coalition it may be necessary for a to pay c some compensation. This could be a problem faced by the governments of Eastland and Westria in their domestic water policies, if we think of a government as in each case a coalition of interest groups within the country. It is common to assume *transferable utility*,[5] which means that a simple transfer of some of c's winnings to a can fully compensate a. This could be true, for example, if all payoffs are in money and the players' "utility functions" are proportional to their money payments, including game payoffs and side payments. This simplifying assumption is often called TU.

Assuming TU, all that matters is the total payoff to the group $\{a, b, c\}$. Therefore, it is common in cooperative game theory to ignore all the details and to focus on the total values the various coalitions can obtain, and assign to each coalition a number expressing that value. It is commonly called the value of the coalition. (Of course, the value of $\varnothing$ is zero.) This assignment is called a "characteristic function" in mathematical set theory and is sometimes called the "coalition function" in cooperative game theory.

Rousseau's tale of the Stag Hunt has given rise to a widely used example in the theory of noncooperative games. Here, though, we will use it as an example of a cooperative game. Rousseau's ideas are an important root of modern political theory and the Stag Hunt example is something of a paradigm of collective action. (For representative treatments see Gardner, 2003, pp. 115–18; Osborne, 2004, pp. 20–21.) Rousseau writes in part 2 of the *Discourse on Inequality* (1754, G. D. H. Cole translation):

> In this manner, men may have insensibly acquired some gross ideas of mutual undertakings, and of the advantages of fulfilling them: that is, just so far as their present and apparent interest was concerned: for they were perfect strangers to foresight, and were so far from troubling themselves about the distant future, that they hardly thought of the morrow. If a deer was to be taken, every one saw that, in order to succeed, he must abide faithfully by his post: but if a hare happened to come within the reach of any one of them, it is not to be doubted that he pursued it without scruple, and, having seized his prey, cared very little, if by so doing he caused his companions to miss theirs. It is easy to understand that such intercourse would not require a language much more refined than that of rooks or monkeys, who associate together for much the same purpose.

It is not clear that Rousseau has anything in mind that corresponds to the rational-action analysis typical of game theory. Rather, he is concerned with lack of foresight among beings not yet human enough to have language! However, the behavior described is not necessarily irrational.

It seems that Rousseau had something like the following hunting technique in mind: the individual hunters act as beaters and spearmen, driving the stag into a narrow defile where it can no longer escape their spears. While they are beating, it may be that one spies a rabbit and, to pursue it, abandons his post, so that the stag escapes through the gap so created. The individual thus forfeits his own share of the stag (while depriving the others of theirs as well) but he may reason that if he does not pursue the rabbit, some of the other beaters will do so, and thus the stag will get away anyway, so that he has nothing to lose by pursuing the rabbit.

It is rather odd, nevertheless, to treat the Stag Hunt as a noncooperative game. In order to pursue a stag, it is necessary for the two (or more) hunters to form a coalition for the purpose, and if they obtain a stag the coalition acquires a single large source of meat, not separate payoffs to different hunters. The hunters must have some agreement as to how the meat will be divided, and the individual payoffs can be based only on the agreement.

Let us instead treat the Stag Hunt game as a cooperative game, and to better illustrate the cooperative-game concepts, let it be a three-person game. As before we suppose that a stag can be taken only if all (three) hunters collaborate in pursuing the stag, that anyone can catch a rabbit, that a rabbit is worth one day's supply of meat for a family while the stag can supply 10 family-days of meat to be divided among the three hunters.

Table 2.4   Game 2.4: a cooperative, three-person Stag Hunt

Coalition	Value
{a,b,c}	10
{a,b}	2
{a,c}	2
{b,c}	2
{a}	1
{b}	1
{c}	1

We can then say that any singleton coalition is worth just one (rabbit) family-day of meat, any two-person coalition is worth two (rabbits) family-days of meat, while a three-person grand coalition is worth ten family-days of meat. This gives us the coalition function shown as Table 2.4. The three hunters are indicated as a, b, and c.

A simplifying assumption that is usually made is that the game is "superadditive," that is, a coalition formed by the merger of two or more coalitions will realize a value at least as great as the sum of the values of the coalitions merged. We see this in the Stag Hunt game.

Now let us consider a game of production of a public good,[6] Game 2.5. Once again we will think of a three-person game, since it allows for the formation of some range of coalitions while remaining comparatively simple. For this game, the three agents, again imaginatively called a, b, and c, each begin the game with wealth amounting to 5 units, supposing that the public good is not produced. For simplicity, we suppose that the public good is imperfectly divisible. Specifically, each agent can provide a fixed lump of 1 unit of the public good at a cost of 3, and the strategies will be to provide or not to provide this indivisible lump of public good. Payoffs will be the sum of the initial wealth and double the number of units of the public good provided, minus the individual's cost of providing the public good if he determines to provide it. We now consider three cases:

(a) *A singleton coalition.*   If the public good is produced, the wealth of the singleton coalition (individual agent) that produces it would be $2k+5-3+2=2k+4$, where k is the quantity of the public good produced by other agents. The payoff from not producing is $2k+5$ so that the singleton coalition would be better off not producing it.

(b) *A two-person coalition.*   A two-person coalition can produce one or two units of the public good, and none will be produced by the singleton residual coalition, as we have already seen. If none is produced, the value of the coalition is 10. With one unit the value of the coalition is $10-3+4=11$. With two it is $10-6+8=12$. Thus the two-person coalition would be better off producing two units of the public good. For this example we assume that they do.[7]

(c) *A three-person coalition.* A three-person coalition can produce up to three units of the public good, and similar reasoning leads to the conclusion that it will produce 3 for a value of $15-9+3*6=24$ if the public good is produced, and 15 if not, so the grand coalition would choose to produce three units of the public good.

Notice, though, that the payoff for a singleton coalition also depends on whether the public good is produced. Suppose, on the one hand, that the group is "organized" into three singleton coalitions. Then, as we have just seen, the worth of each is 5. Suppose, on the other hand, that a and b form a two-person coalition and produce two units of the public good. Then the value of $\{c\}$ is $5+4=9$. Evidently, the value of a singleton coalition in this example depends on whether the other two agents in the game form a coalition among themselves, and if so what strategy they jointly choose.

These changes in the value of a coalition, as a result of the formation of another coalition, are called *externalities*[8] in some recent work in game theory (for example, Carraro, 2003). When the coalition function is taken as the only important information about the game, as so much cooperative game theory has done, this amounts to the unstated assumption that externalities are unimportant.[9] This, too, seems quite problematic for economic applications. As early as 1963 Thrall and Lucas proposed a more complex way of assigning values to coalitions, the *partition function*, that allows for externalities in this broad sense.

A partition is a common construct of set theory. A *partition* $\mathcal{P}$ of set N is a set of subsets of N with two properties: the subsets do not overlap and every member of N is in one or another of the subsets that are members of $\mathcal{P}$. A second partition $\mathcal{Q}$ is a *refinement* of $\mathcal{P}$ if it breaks down the subsets that are members of $\mathcal{P}$ into smaller subsets. These can be applied to sets of any kind, from brands and models of automobiles to classes of probability distributions over the unit interval. Thus, for example, the set of all automobiles is partitioned when we identify them according to brand, and since each model (such as the Chevrolet Impala or the Subaru Outback) is produced by only one brand, the identification of automobiles by model is a refinement of the identification by brands. The *partition function* then is a function from coalitions and the partitions of which the coalitions are members to the value of the coalition, where the coalition can take different

Table 2.5   Game 2.5: a partition function for the three-person public goods game

	Partitions	Values
1	$\{a, b, c\}$	24
2	$\{a, b\}, \{c\}$	12, 9
3	$\{a, c\}, \{b\}$	12, 9
4	$\{a\}, \{b, c\}$	9, 12
5	$\{a\}, \{b\}, \{c\}$	5, 5, 5

values in different partitions. An *imbedded coalition* is a pair consisting of a partition $\mathcal{P}$ and a coalition C that is a member of the partition. Thus the partition function assigns values to imbedded coalitions. (For more detail see McCain, 2013, Chapter 3.)

A partition function for the public goods game, as discussed above, is shown in Table 2.5. Once again, this game is superadditive. Suppose, nevertheless, that we focus on one of the lines other than the grand coalition, and ask for a "cooperative solution," that is, an assignment of values to the individual agents in the coalitions in that partition. An example would be the middle line, $\{a, c\}, \{b\}$, where the problem would be to determine the allocation of the coalition value of 11 between a and c, consistently with the assumptions of cooperative game theory. The partition $\{a, c\}, \{b\}$ would be called a *coalition structure*, and the solution would be for that coalition structure.

As we observed, a partition is a general concept of set theory. It will sometimes be helpful to think in terms of partitions of other kinds of sets. For example, any decision can be thought of as creating a partition among the different possible outcomes of a game. In Selten's Horse Game, for example, Firm A's decision creates a partition of the four possible outcomes into two sets of two: $\{(0,0,0), (3,2,2)\}$ and $\{(0,1,1), (4,4,0)\}$. Similarly, chance events partition all possibilities into different subsets that may follow from different realizations of the chance event.

2.6   "IMPERFECT RECALL"

In his classic paper that founded the literature on games in extensive form, many of Kuhn's (1997, pp. 46–68) results were limited to the case of "perfect recall." Kuhn's example of "imperfect recall" was the game of bridge. In bridge, two partnerships play against one another, and each partnership receives a single score. Thus, Kuhn argued, the partnership is a single player, but the player has multiple "agents" each of whom may be unaware of the strategic commitments of the other. Many of the conventions of bridge play are designed to address this difficulty.

Nontrivial coalitions are compounds of multiple agents, and when we observe coalitions interacting competitively against one another, they are compound players such as Kuhn described. Nevertheless, there seems to be no literature whatever on cooperative solutions for games of "imperfect recall." The popular view is that cooperative game theory has been less influential in economics since the early period because noncooperative game theory is better able to deal with informational problems (Weintraub, 1992, p. 7.) To address these problems in cooperative game theory would require a discussion of games of "imperfect recall." Even for superadditive games in coalition function form, such as exchange games, "imperfect recall" can have crucial implications, points that will be discussed in Chapter 9.

2.7 NON-NUMERICAL OBJECTIVES

So far we have assumed that the consequence of choosing a particular strategy, while others choose their strategies, is a numerical payoff to each player. In many cases the numbers we have chosen have been based on very rough reasoning, a measure of conjecture, and no more. Numerical payoffs lend themselves to transferable utility games, in which side payments take a central role; and side payments are very common in the real world. Salaries, wages, interest payments, dividends and taxes are all instances of side payments. While side payments can be part of a game without transferable utility, in TU games the determination of side payments can be treated explicitly rather than being left in the background. Numerical payoffs are almost always assumed in the theory of noncooperative games. Moreover, von Neumann and Morgenstern thought very carefully about how the numbers should be established, and their *von-Neumann–Morgenstern utility theory* has not been improved upon. We have not gone into that, because it is not central to our purpose. However, it will be worthwhile to sketch how game theory might be done without numerical payoffs.

We may, then, define a game in a more general way. To be concrete, here is an example drawn from World War II. The successful Allied invasion of Normandy followed a massive and remarkably successful attempt to deceive the Germans into believing that the Allied invasion would be elsewhere. By this time, Germany could not have achieved a victory and hoped by defeating an Allied invasion to bring the war to a stalemate that

Table 2.6   Game 2.6: the Normandy invasion game

		Germany	
		Defend Normandy	Defend Calais
Allies	Attack Normandy	Stalemate	Victory
	Attack Calais	Victory	Stalemate

would have allowed the Nazi regime to survive. The Germans were led to believe that the invasion would be at Calais rather than Normandy, and this continued after the invasion itself, leading the Germans to believe that this invasion was a feint and the real, larger invasion was yet to come at Calais. With some simplification – well, with a lot of simplification – we can see this as a two-by-two game, as shown in Table 2.6. By this stage in the war, each country had two alternatives to choose from and there were just two outcomes: Allied victory or stalemate. These are shown in place of payoffs in Table 2.6.

How are we to attach numbers to these outcomes? If we assign a value of zero for both sides to a stalemate, and one for the Allies and minus one for the Germans in case of Allied victory, this seems to represent the case adequately, within the limits of the information that we are given. We can do no better than that.

The *outcome* of a game, then, is a complicated object that may have many numerical dimensions or none, and for some dimensions may be Boolean, (that is, either-or states as with the Normandy invasion) or real numbers, or other number forms, perhaps in combination. All that we insist on is that there is a set of outcomes, comprising no less than two, and that the players are allowed to disagree as to which one they would wish to have realized. For games of exchange, an outcome is an allocation of goods, services and resources among the agents in the market, and has dimensions of all goods and services for each agent. For another example, think of a political coalition in a parliamentary government as in many countries of the continent of Europe. The formation of a governing coalition is a complex game comprising various elections and negotiation among parties following the election. The outcome of such a game is a government program, which may (depending on the strategies of the contending parties) include more or less subsidy to agriculture, an aggressive or defensive foreign policy, support or nonsupport for an established church and its clergy. Even if we cannot assign utility numbers to outcomes, we suppose that players have preferences over them and can evaluate them in relative terms.

This more general approach to representing games has been used primarily in cooperative game theory, especially where (realistically) there may be difficulties making side payments. In that case, we think in terms of non-transferable utility (NTU) games. This approach is little used in noncooperative game theory, but can be useful in thinking through practical problems even when the approach is noncooperative. As we begin to frame a policy problem as a game theory model, it will often be more natural to begin by asking what different outcomes are possible, and thinking of the strategies as determining outcomes, and only then to translate the outcomes into payoff numbers. Recall the Water Game, in which (hypothetically) the payoff numbers were derived from cost–benefit analysis. But the economists who conducted the cost–benefit analysis will have begun with some qualitative listing of the outcomes of the water policies of the two countries: ample supplies of drinking water in one country or another, more or less depletion of water for irrigation, silting and loss or costly maintenance of navigable channels.

2.8   CHAPTER SUMMARY

Since game theory is interactive decision theory, representing a social arrangement as a "game" means representing it as an interactive decision. There are two major ways of doing this for noncooperative game theory. The most commonly used is the game in strategic normal form, essentially a table or function from the strategy choices of all players to their payoffs. For this purpose, it is important to represent the choice of behavior strategies as contingent on any information the agent may have at the time a decision is made. The second way of representing games for noncooperative game theory is the game in extensive form, that is, a decision tree in which the different "players in the game" govern different decision nodes. With information sets, this allows a much more explicit representation of the information available in the game, and is often used when information is important. For cooperative games, the game is usually represented by a coalition or characteristic function, assigning a total value to each coalition that might form. This is usually linked to the assumptions of transferable utility and superadditivity. For practical purposes, however, the coalition function and superadditivity are often not helpful as they assume away the very problems we want to analyze. Accordingly, we may instead represent the game as a partition function (in which the value of each coalition depends on the other coalitions that form) and attempt to determine which partition, or coalition structure, is likely to be observed. Despite the long history of this approach, however, the theory remains unsettled. As a rule, we will think in terms of numerical payoffs and transferable utility, recognizing that this is at best an approximation to reality, but usually, we may hope, a good one.

NOTES

1. My students feel that a better term would be "ignorance set," since it is expressive of ignorance rather than information. Nevertheless, of course, we will use the conventional terminology.
2. It is common, as a matter of convention, to record first the payoff to the player to the left, but in this book the order will routinely be indicated in the upper left space. I am indebted to my colleague Richard Hamilton, MD, for that presentation.
3. This follows Kuhn's terminology. We should observe, though, that "behavior strategies" in this sense have nothing to do with the perspective of behavioral science or "behavioral game theory," in that the behavioral strategies may be supposed to be chosen with rationality that is flawless, regardless of experimental evidence of the limited cognitive capacity of real human beings to make such choices.
4. This specific example was included because of its importance in the history of game theory. If the example of patent licensing seems a little stretched, it is because the original example had no such application in mind – and indeed no application of this example has ever been offered before, to the best of my knowledge! The representation of the patent licensing problem could probably be improved; but the general principle – that a decision maker may not know the sequence of decisions by others that has brought his own decision about – probably is not uncommon in reality.
5. This phrase recognizes that the ultimate benefits of economic activity are subjective, that is, in the economist's language, "utility," but assumes that utility is nearly enough proportional to money that money side payments will assure that everybody in a coalition benefits on net from the coalition's activity.
6. When written in this way, with the indefinite article – "*a* public good" – this term makes reference to a concept systematically discussed by Paul Samuelson (1954). With the definite article – "*the* public good" – the meaning is of course much broader and the usage much older. Unfortunately, the two meanings are often confused. A (Samuelsonian) public good is defined by its technical conditions of production and offering: the incremental cost of adding one user is always zero, and there is no practical way to make payment a condition of use. Thus provision at a zero price is (on the one hand) unavoidable and (on the other hand) efficient, since the marginal cost is also zero. In particular, free provision from general tax revenues or on a philanthropic basis is efficient.
7. Most cooperative game theory follows von Neumann and Morgenstern in assuming (on the contrary) that coalition values are the worst that an opposed coalition can bring about. For more detail see McCain (2013, Chapter 1), and for an alternative interpretation Chapter 8, section 8.4.
8. The term "externalities" is used more narrowly in economic theory. When a cartel is formed and thus imposes costs on customers, this would have been called a "pecuniary externality" in, for example, Scitovsky (1954); but modern economic theory does not regard "pecuniary externalities" as externalities.
9. There is a modest literature on externalities in cooperative games in coalition function form. A key paper is Shapley and Shubik (1969).

3.   A brief interpretive history of game theory

Game theory has, of course, a prehistory; but much as we can say that economics (as a distinct field of study) began with Adam Smith's (1776/1994)[1] *Wealth of Nations*, so we can say that game theory began with von Neumann and Morgenstern's (1944/2004) book, *The Theory of Games and Economic Behavior*. Accordingly we shall pause only briefly over the prehistory. In 1913, Zermelo had initiated the mathematical literature on analysis of games. Borel had written important papers that seem to have influenced von Neumann (Poundstone, 1992, pp. 41–2). With a presentation in 1926 and publication of the paper in 1928, von Neumann (1928/1959) had set out many of the themes to recur in his book with Morgenstern, to which we will return. Aumann and Maschler (1985) find a cooperative solution concept in the Babylonian Talmud. A Korean scholar suggests to me that Sun Tzu should be considered a game theorist.[2] Indeed it is likely that insights of game theory have often occurred to thoughtful people engaged in their own conflicts throughout much of history. See Paul Walker's website for a schematic history of game theory, including several other "prehistoric" contributions.[3]

Morgenstern had used his example of Sherlock Holmes and Moriarty to illustrate problems of interactive decision, relating them to economics and forecasting, in a 1928 book and a 1935 paper (TGEB,[4] pp. 712–14). By the mid-1930s the convergence of his ideas with those of von Neumann had been pointed out to Morgenstern, but Morgenstern was unable to pursue that direction until he had been dismissed from his position in Vienna by the new Nazi regime as "politically unbearable" (TGEB, p. 715).

3.1   THE FOUNDING BOOK

In his 1928 paper von Neumann assumed that games would have numerical payoffs. In the founding book, von Neumann and Morgenstern first take up the nature of the payoff numbers. Assuming that the ultimate benefits of any economic activity are subjective, in the nature of utility, Chapter 1, Section 3 develops a numerical utility concept. This is the von

Neumann–Morgenstern utility index. Given that agents have consistent subjective preferences over risky prospects such as lottery tickets, the index allows utility to be expressed in terms that are consistent with decisions that maximize expected values of the utility index. These decisions will then be rational in terms of the given preferences.

Von Neumann and Morgenstern then take up the character of a solution in game theory. Ideally, the solution is an imputation, that is, it tells us the quantity of "utility" that each participant in the game can expect on the basis of "rational behavior." For each individual, the solution would also constitute a set of rules for rational behavior in any conceivable circumstances. However, this may not always be possible, and in general the solution may constitute just a set of imputations.

In the second chapter von Neumann and Morgenstern address the representation of a game (interactive decision problem) for mathematical analysis. Von Neumann had in 1928 given both of the representations now common for noncooperative games, the extensive and normal form.[5] For a given game in extensive form (sequence of decisions at the successive stages of the game) each *strategy* for a given player is the sequence of decisions made in the successive stages of the game. But a key point is that these decisions are contingent decisions, as discussed in Chapter 2 above. Von Neumann writes (1928/1959, p. 18) "For each possible combination of results of the 'draws' and 'steps' [known to player S_m] . . . it must be specified what S_m's decision . . . is going to be."

In the book, three representations are given. The first is the intuitively appealing idea of a game as a sequence of decisions at the successive stages of the game, some by individual participants in the game, and some by a random mechanism. This can be visualized as a tree diagram (TGEB, p. 65) and corresponds to the game in extensive form as it is discussed in recent game theory. The second, however, is a rather more complex object. Here von Neumann and Morgenstern represent the game as a sequence of partitions. Von Neumann and Morgenstern begin with a set of all outcomes of the game, and decisions and chance events are treated as partitions of this underlying set of outcomes, and later decisions or chance events produce refinements of the partitions generated by earlier decisions and events. (For this discussion, an "outcome" is simply a list of the payoffs to all players.)

Von Neumann and Morgenstern then discuss the meaning of a *strategy* for a game specified in this way: "Imagine now that each player . . . instead of making each decision as the necessity for it arises, makes up his mind in advance for all possible contingencies . . . We call such a plan a *strategy*" (TGEB, p. 79). Once strategies have been defined in this way, as contingency plans, it is possible to return to the "very simplest description,"

(TGEB, p. 79), that is, the game in strategic normal form. This is the third and crucial representation in von Neumann and Morgenstern's Chapter 2.

The significance of the formalization in terms of partitions is that it clearly establishes the link between the game in extensive form and the game in strategic normal form. The treatment of information in the game expressed as a sequence of partitions is cumbersome, but it disappears in the game in strategic normal form. This powerful and brilliant simplification undoubtedly accounts for much of the impact of game theory on economics and on other fields in which strategy or interactive decisions are important.

In the most famous and definitive section of the book, Chapter 3, von Neumann and Morgenstern address two-person, zero-sum games, deriving the maximin solution in mixed strategies that von Neumann had already discussed in the 1928 paper. A brief digression on mixed strategies is in order. The set of strategies (sequences of contingent choices) already discussed are designated as "pure strategies." In a mixed strategy equilibrium the players may "randomize" their choices of strategies; that is, they may choose among the finite set of "pure" strategies according to a probability distribution. It is the probability distribution that is adapted so that the player balances the advantage of choosing a particular strategy against the danger of having her strategy "found out" by the opponent (to use a phrase that recurs in *The Theory of Games and Economic Behavior*).

The remainder of *The Theory of Games and Economic Behavior* was devoted to generalization of that model to the more general case of n-person, nonconstant-sum "games." To do this von Neumann and Morgenstern adopted a research strategy that has been highly productive in mathematics, literally for thousands of years. The strategy is to make the solution to the simple case, with some appropriate transformation or extension, the solution to the more complex case.

In the 1928 paper, von Neumann had made the zero-sum restriction part of the definition of the game. The introduction of nonconstant-sum games was an important contribution of *The Theory of Games and Economic Behavior*, and it was clearly essential in order that the theory be applicable to "economic behavior" and to the social sciences in general. Nevertheless, in 1928 von Neumann sketched some essentials of a theory of n-person zero-sum games, with some of the difficulties to be encountered. Even for three-person games, von Neumann admitted no solution that would not allow for coalitions (von Neumann 1928/1959, p. 33). Coalitions are also central to the discussion subsequent to Chapter 3 of *The Theory of Games and Economic Behavior*. This is seen as being all the more crucial for applications to "economic behavior;" as von Neumann and Morgenstern write (TGEB, p. 15) "the great number of participants [in potentially competitive markets] may not become effective; the decisive exchanges may take place between large 'coalitions,' few in number." In a footnote, they elaborate: these coalitions may include "trade unions, consumers' cooperatives, industry cartels, and conceivably some organizations more in the political sphere."

Thus, in the remainder of *The Theory of Games and Economic Behavior*, they develop theories of n-person games first with and then without the zero-sum restriction. In each case, they proceed step by step, with detailed analyses of three-person games and some other small-n cases both as preliminary case studies and in reconsideration. For this purpose they adopt the transferable utility assumption and define the characteristic (or coalition) function as a fourth representation of the game. This step, representation of a coalition by the total value it can realize, reflects the fact that side payments may be necessary to form some coalitions and the "transferable utility" assumption implies that they can be made costlessly. For n-person zero-sum games, they argue that the game will always be resolved to a confrontation between two coalitions with absolutely opposed interest. "Since we have an exact concept of 'value' (of a play) for the zero-sum two-person game, we can also attribute a 'value' to any given group of players, provided that it is opposed by coalition of all the other players" (TGEB, p. 238).

For games represented in coalition function form, von Neumann and Morgenstern again proceed largely as von Neumann had done in his 1928 paper. A candidate solution is an "imputation," that is, an assignment to each player of the amount he can expect to receive, with the total of the amounts for each coalition limited by the value of that coalition. They then define a dominance relation on imputations, as follows: one imputation dominates another if there is a set of players who can form a coalition and force the second imputation, and increase their payoffs as a result. Unfortunately (as von Neumann had noted in the 1928 paper) this relation is not transitive. The solution then consists of all imputations such that (1) an imputation in the solution is not dominated by any other imputation in the solution, and (2) every imputation outside the solution is dominated by at least one imputation within the solution. Dominance cycles are a possibility, since an imputation in the solution can be dominated by an imputation outside the solution. There may be many imputations in the solution, and moreover there may be many solutions, and this multiplicity is recognized as a shortcoming of their solution concept (TGEB, pp. 264–6).

The final step is to extend the solutions to n-person nonconstant-sum games. To take this step, von Neumann and Morgenstern construct an $n+1$ 1-person game $\overline{\Gamma}$ corresponding to the n-person game Γ. In $\overline{\Gamma}$, the $n+1$ 1st player is simply a fictitious player whose payoff is the negative of the sum of the payoffs of all the others (TGEB, pp. 505–6). The solution to the zero-sum game $\bar{\Gamma}$ will then be the solution to Γ. However, this requires some modifications, in that the fictitious player controls no strategies and can join no coalitions and make no side payments. Once these elements are included in the "rules of the game," the analysis of n-person constant-sum games is recapitulated. Since the game is superadditive, one may presume that the grand coalition of all actual players will form to exploit the fictitious player (nature?) most effectively, but it remains to determine how the overall gain will be distributed. To set limits on this, once again, the game is expressed in coalition function terms. In the absence of the grand coalition, the situation will again resolve itself to an opposition between two coalitions of actual players. (Just two, since any partition into three or more coalitions will be unstable. With superadditivity, the opposition will be better able to defend themselves by merging into a single oppositional coalition.) Von Neumann and Morgenstern then again apply the maximin theorem to assign a value to each coalition. To do this, they have to assume that a coalition S will face a counter-coalition of the remainder of the actual players in Γ, called $-S$, who will inflict maximum harm on S even at cost to themselves. This procedure is called "the assurance principle" and the value obtained is "the assurance value," as it is the largest value the coalition can assure themselves of in all circumstances. Von Neumann and Morgenstern concede that "the desire of the coalition $-S$ to harm its opponent, the coalition S, is by no means obvious. Indeed the natural wish of the coalition $-S$ should be not so much to decrease the expectation . . . of the coalition S as to increase its own expectation . . ." (TGEB, p. 540). However, the assurance principle is nevertheless assumed, as inflicting harm is seen as a threat strategy by which the group in $-S$ would hope to influence the others and increase their imputation in the grand coalition that will ultimately form (TGEB, pp. 541, 559). With these qualifications, the dominance relation is again used and a solution set of imputations, perhaps quite a large set, is derived.

As with constant-sum games, solutions of general games could include multiple imputations and there could be many solutions; moreover, it was not known whether every game had a solution. (Lucas, 1968, later demonstrated that it is not true that every game has a solution.) Further, von Neumann and Morgenstern concede that their analysis of n-person nonconstant-sum (general) games depends very crucially on the assurance principle (TGEB, p. 559) and, after all, "$\bar{\Gamma}$ is merely a 'working hypothesis'" (TGEB, p. 540) based not on "a purely mathematical analysis [but] more in the nature of plausibility arguments" (TGEB, p. 506) and to be vindicated, if at all, by its success in applications (TGEB, p. 542). The relative lack of applications of the von Neumann–Morgenstern solution sets in recent game theory suggests that the working hypothesis was inadequate. This is hardly surprising in the founding work: at that time there existed no theory to expand the implications of the possibility that "the natural wish of the coalition –S should be . . . to increase its own expectation" in a nonconstant-sum game (TGEB, p. 540). Nash's (1950a, 1951) equilibrium theory would address that, but was of course not available to von Neumann and Morgenstern; and despite the emergence of the Nash equilibrium theory, writing on cooperative game theory tends to assign values via the assurance principle even today (Telser, 1978; Peleg and Sudhölter, 2003; Forgo et al., 1999). In any case, the founding book of game theory had founded not one but two important streams of research: with the theory of two-person, zero-sum games it founded noncooperative game theory, and with the theory of n-person nonconstant-sum games it founded cooperative game theory, providing the concepts and solutions in particular cases without which neither would have grown.

3.2   THE DICHOTOMY OF COOPERATIVE AND NONCOOPERATIVE GAMES

The appearance of *The Theory of Games and Economic Behavior* caused a great deal of interest, of course, especially at Princeton. Game theory was promptly taken up for defense research by the new RAND corporation, and some of the book reviews published were themselves important contributions. Nevertheless, such a path-breaking work required a few years for absorption, and the next important advance occurred in 1950 as John Nash reported (1950a) an equilibrium concept for n-person nonconstant-sum games. Nash would expand this (1951) into an explicit theory of noncooperative games. It is important that Nash's equilibrium solution is identical to that of von Neumann and Morgenstern in the case of two-person, zero-sum games. Thus it is again an instance of the classical research strategy of mathematics, making the solution for a simple case, with some appropriate transformation or extension, the solution to the more complex case. Nash's solution differs for all games with three or more players and all nonconstant-sum games because it does not allow for coalitions based on enforceable agreements.

But Nash also (1950b, 1953) made an important contribution to the theory of cooperative games, in the form of an axiomatic theory of bargaining. This will be discussed in Chapter 10. Further, Nash provides a model for the development of the theory of cooperative games in general. In Nash (1951; CGT,[6] 1997, p. 26) he writes

"One proceeds by constructing a model of pre-play negotiation so that the steps of the negotiation become moves in a larger noncooperative game . . .

describing the total situation. . . . if values are obtained they are taken as the values of the cooperative game. Thus the problem of analyzing a cooperative game becomes the problem of obtaining a suitable, and convincing, noncooperative model of the negotiation."

This reduction of cooperative game theory to noncooperative game theory is the *Nash program* (Serrano, 2003).

In this series of papers, Nash not only extended both noncooperative and cooperative game theory, but in addition originated the distinction between the two. No such distinction exists in von Neumann and Morgenstern. The idea that the same game might have alternative solutions, cooperative and noncooperative, with the first applicable only in case enforceable agreements can be made, originates with Nash and is one of the most influential ideas of game theory.

The first experimental study in game theory probably took place in 1949 at the RAND corporation. Merrill Flood involved two secretaries in an experiment that roughly anticipated the Ultimatum Game and the Dictator Game, with surprising results. This was followed in January, 1950, with a formal experiment in noncooperative games. Melvin Dresher collaborated and the subjects were John Williams and Armen Alchian, respectively a mathematician and an economist. While the experiment is of considerable interest in itself, its greatest impact probably was indirect. Alfred Tucker observed the experiment and, in May 1950, addressing a group at Stanford University, originated the Prisoner's Dilemma example. The Prisoner's Dilemma is a symmetrical modification of the game in the Flood–Dresher experiment. (This account follows Poundstone, 1992, Ch. 6.) The Prisoner's Dilemma outcome is a particular case of Nash equilibrium, but a simple and compelling instance in which individual self-regarding action makes both parties worse off than they might otherwise be. As such, it was to have enormous impact and this has been one very important reason for the predominance of noncooperative approaches in game theory in the later twentieth century.

The game theory research of the 1940s was reflected in 1950 by the first volume of *Contributions to the Theory of Games*, edited by Kuhn and Tucker, as number 24 of the *Annals of Mathematics Studies*. Many of these studies are extensions of and computational approaches to the maximin theorem; games with an infinite number of strategies are seen as the research frontier (Kuhn and Tucker, 1950, p. x). Some interesting new developments are found in the collaboration of Nash and Shapley on a simplified three-person poker game. At p. 109 we see what seems to be the first elimination of dominated strategies in the solution of a noncooperative game. (While the other papers in this volume were also important contributions, brevity will require selectivity from this point on.)

Table 3.1   Game 3.1: McKinsey's game in strategic normal form

Payoffs: Player 1, Player 2	Player 2	
	A	B
Player 1	0, −1000	10,0

Table 3.2   Game 3.1: McKinsey's game in coalition function form

v{1}	0
v{2}	0
v{1,2}	10

McKinsey published in 1952 what seems to have been the first textbook of game theory. One important novelty in this book (Luce and Raiffa, 1957, p. 190; Tucker and Luce, 1959, p. 2) is the beginning of serious criticism of the representation of cooperative games in coalition function form. McKinsey writes (1952, p. 351) "von Neumann's whole theory of games is based on the notion of the characteristic function. This implies that if two games have identical characteristic functions, then they will have the same solutions. It is, to say the least, debatable, however, whether this is satisfactory from the point of view of intuition." He considers (1952, pp. 351–2) a two-person game in which Player 1 has only one strategy (no alternatives) and Player 2 has two. The game is shown by Table 3.1. As we can see it is highly asymmetrical, and intuition suggests that Player 1, despite his lack of alternatives, is in the better position. Player 2 can avoid a very large loss only by choosing strategy B, which grants a payoff of 10 to Player 1. Nevertheless, when the game is expressed in coalition function form, it is symmetrical. For notice that the least payment Player 1 can assure himself of is 0 (for although Player 1 can take no action his payoff cannot be less than zero in any case). The least payoff of which Player 2 can assure himself, by choosing strategy B, is zero. Therefore the game in coalition function form is as shown in Table 3.2.

Thus, the coalition (characteristic) function form is symmetrical, failing to capture the most important aspect of the game in strategic normal form. Moreover, when we consider v{1} = 0 as reflecting a threat by Player 2, it is not very plausible. To reduce Player 1 to the payoff of 0, Player 2 must take a loss of 1000. According to the assurance principle, this is what Player 2 would do. Is it likely that Player 2 would make such a threat or (more importantly) that Player 1 would find it credible if he did?

In 1952, Shapley and Shubik presented an analysis of cooperative games without the assumption of transferable and linear utility (TU) that von Neumann and Morgenstern had made. This was in a conference of the Econometric Society at East Lansing, Michigan, and the paper is apparently available only in the form of the abstract published in *Econometrica*. Nevertheless it deserves mention here. Shapley and Shubik assume that preferences can be indicated by a numerical index that would not be transferable nor interpersonally comparable, but which attaches higher numbers to more preferable alternatives. Corresponding to any outcome or probability mixture of outcomes would be a vector of utility indices for the N players in the game. A coalition S is then "effective" for utility index vector $\mathbf{x}$ if there is a joint strategy or mixture for S that will assure them of at least the utility indices in $\mathbf{x}$ regardless of the strategies chosen by the players not in S. This definition of effectiveness is equivalent to von Neumann and Morgenstern's assignment of coalition values via the maximin operation, that is, the assurance principle. They then define dominance and solution in terms of effectiveness, otherwise following the example of von Neumann and Morgenstern.

The second volume of *Contributions to the Theory of Games* (Kuhn and Tucker, 1953) contained two very important new contributions that would be republished in the collection *Classics in Game Theory* (Kuhn, 1997). The first of these was Kuhn's "Extensive Games and the Problem of Information" (CGT, pp. 46–68). Here Kuhn returned to the representation of a game as a series of partitions of the set of all outcomes, but defined the partition in a different and more general way that allowed for a treatment of the information available to a player at a particular play in a way that is at once more compact and general. The sets that make up Kuhn's "information partition" are the "information sets." Adopting Nash's equilibrium concept as a generalization of the maximin solution, Kuhn proves that all games of perfect information have equilibria in pure strategies, an extension of the theorem of Zermelo and von Neumann. Kuhn also supplied a geometric visualization of extensive games and their information conditions that has become standard (note CGT, p. 64). Kuhn defines subgames in the way that has also become standard (CGT, p. 56).

Kuhn's formalization, unlike that of von Neumann and Morgenstern, extends to games in which a player may not be aware of the number of plays that have already taken place (such as Selten's "Horse;" also note CGT, p. 52). It also includes games in which a player is represented in different plays by different "agents" some of whom may be unaware of previous moves made by other "agents" of the same player, that is, games of "imperfect recall" (CGT, p. 65). Kuhn advocates "behavior strategies," that is, local randomization at each step of decision, making use of the information available at that point, rather than the contingent pure strategies of von Neumann and Morgenstern. Kuhn points out considerable computational advantages of behavior strategies, in that rational decisions need not be computed for decision points that will never be reached.

As we noted, Kuhn took Nash's equilibrium as his concept of solution, extending that concept by assuming that at each decision point, the behavior strategies chosen would be *local* best responses given the information available at that point. As a rule (he stressed) these would be randomized strategies. On that basis, he proved (1) that every sequence of behavior strategies chosen in this way would correspond to at least one contingent strategy, (2) every contingent strategy leading to the same payoff outcomes would be identical to the equilibrial sequence of behavior strategies on the information sets actually reached, although there might be many such contingent strategies with different decisions on "irrelevant" information sets not actually reached in equilibrial play, and (3) if the game has "perfect recall," then every such sequence of equilibrial behavior strategies corresponds to a Nash equilibrium of the original game. This is sometimes expressed by the phrase "behavior strategies suffice" and probably accounts for the neglect of the distinction of behavior and contingent strategies in much subsequent work in noncooperative game theory. This will be reconsidered in Chapter 9.

The other contribution from volume 2 that must be mentioned here is Shapley's value theory, a solution concept for n-person cooperative games (CGT, pp. 69–79). The solution is a value function which, for a given game, assigns a value to each player that is the player's expected payoff from participating in the game. It will be discussed in Chapter 8, at Section 8.2.2. Shapley and Shubik (1954) applied the value theory as an index of power in political organizations.

Although it was not to be published until volume 4 of *Contributions to the Theory of Games*, (CGT4, pp. 47–85),[7] Gillies had by this time (1953) developed the concept of the core of a cooperative game and presented it in his doctoral dissertation. The core, like the von Neumann and Morgenstern solution set but unlike the Shapley value, may include many imputations or may be null; that is to say, there may be no imputations that meet the conditions for membership in the core. Thus, with Shapley's value theory, there were three distinct concepts of solution of cooperative games, an *embarras de richesses* that was only to become more pronounced.

3.3 GAME THEORY AS DECISION THEORY

In 1957 Luce and Raiffa published *Games and Decisions*, the first book-length work of research on game theory after von Neumann and

Morgenstern. Much of the ground covered was that of von Neumann and Morgenstern, as Luce and Raiffa regarded that work as still the canon for game theory. However, Luce and Raiffa aimed at a more accessible, less mathematical presentation, and they incorporated a number of advances made over 1944–57, including Nash equilibrium in noncooperative games (GD,[8] Chapter 5), linear programming (GD, p. 17, appendix 5), and Kuhn's formulation of games in extensive form with information sets (GD, p. 42), although they continue to characterize pure strategies and games in normal form in a way consistent with von Neumann and Morgenstern, as a series of contingent decisions with a decision chosen in advance at each information set (GD, p. 51). In cooperative game theory they discuss McKinsey's criticism of the characteristic function (GD, p. 190), the core solution concept (GD, pp. 192–6), Vickrey's then unpublished attempt to introduce farsightedness into von Neumann–Morgenstern solution theory, and also incorporate Nash's bargaining theory and the Shapley value, with some criticisms of them. An original point is that Luce and Raiffa treat these as alternative arbitration schemes (GD, Chapter 6, parts 4–10). Among the advances in this book were a very early discussion[9] of repeated play in the Prisoner's Dilemma (GD, p. 99), including the concept of the unraveling of a cooperative agreement from the last period forward, what seems to have been the first discussion of correlated equilibria in noncooperative games (GD, pp. 116–19), and a discussion of cooperative arbitration schemes in case interpersonal comparisons of utility may be made (GD, Chapter 6, parts 10–11). They discuss cooperative games in strategic normal form (GD, Chapter 7), present their own solution concept for cooperative games, ψ–stability (GD, Chapter 10) and incorporate the work of Savage and its sequelae (GD, Chapter 13) on decisions under uncertainty and of Arrow on collective decisions, along with some discussion of elections, into their game-theoretic framework. And this is not a comprehensive summary!

Luce and Raiffa's ψ–stability deserves some further comment. As they observe, (GD, p. 191), the characteristic function provides very little information about the game. They propose a more informative beginning point: the characteristic function along with a "boundary condition" in the form of a function, ψ, from partitions into sets. If τ is the current partition and S is an element of $\psi(\tau)$, then S is a group of players, not a coalition in τ, that can form and enforce a new partition if its members should choose to do so. If S is not in $\psi(\tau)$, then it is not capable of upsetting the existing partition however much it might have to gain. At the same time, the ψ–stable solution is a partition of the population into coalitions – a coalition structure – as well as a set of imputations. This seems to be the first attempt to construct a theory that would explain a stable "coalition structure" other than the grand coalition.

During the 1950s, the theory of differential games (games in which strategies evolve over continuous time) was developing rapidly, and several papers in *Contributions to the Theory of Games*, volume 3 (Dresher et al., 1957) and in *Advances in Game Theory* (Dresher et al., 1964) focused on this theory. However, these will not be important in the chapters that follow.

Volume 4 of *Contributions to the Theory of Games* (Tucker and Luce, 1959) was focused on n-person games (CTG4, p. 1) and reflects especially the search for alternatives to the von Neumann and Morgenstern solution set, especially the definition of the characteristic (coalition) function in terms of the assurance principle (CTG4, p. 2). Papers by Shapley (CTG4, pp. 145–62) and by Shubik (CTG4, pp. 267–78) applied cooperative game theory to market exchange. Shubik had throughout the 1950s published a number of contributions relating game theory to economics (and to some extent to management and political science). His paper introduced the idea that the core of a market game would correspond to Edgeworth's market theory, an idea that was to dominate applications of cooperative game theory (and, arguably, of game theory in general) to economics in the decade to follow. Vickrey (CTG4, pp. 213–46) proposed to modify the von Neumann–Morgenstern theory by taking into account that some dominance relations among imputations might be shortsighted (though he did not use that terminology explicitly). Aumann (CTG4, pp. 287–324) introduces the supergame as follows. Consider an infinite sequence of repetitive plays of the noncooperative game Γ. Suppose that the players adopt rules to determine their choice of strategies in each play of Γ, where the rules may be conditioned on the other agent's past play. The supergame is the *noncooperative* game of choosing rules by which Γ will be played. Aumann then identifies the cooperative solution of Γ with a noncooperative (strong Nash) equilibrium of the supergame. Many of the powerful developments in the theory of repeated play over the following fifty years are suggested in this paper. Harsanyi (CTG4, pp. 325–56) proposed a generalization of Nash's bargaining theory to n-person cooperative games. Kemeny (CTG4, pp. 397–406) sounds the call for more informative priors: "While I agree that with the information usually given for n-person games no more can be said [than the von Neumann Morgenstern solutions], it seems to me that we must ask for more information" (CTG4, p. 398). But the editors respond, "The difficult question, then, is what more to assume" (CTG4, 1959, p. 11). Kemeny adds an index of the bargaining power of each agent and builds his (relatively informal) solution concept around that.

Thomas Schelling's *Strategy of Conflict* appeared in 1960, though some chapters had been published earlier. A major motivating factor in the book is the game-theoretic analysis of the repressed conflict between the United States and the Soviet Union, which probably was then near its peak of intensity. The book is usually remembered for the focal equilibrium concept (to which we will return) but more systematically the book explores the insight that in noncooperative games "the power to constrain an adversary may depend on the power to bind oneself" (SC,[10] p. 22) with a voluntary sacrifice of freedom of action. This idea is inherent in noncooperative games. Nash had written (Nash, 1953, p. 130) "Supposing A and B to be rational beings, it is essential for the success of the threat that A be *compelled to carry out his threat* T if B fails to comply. Otherwise it will have little meaning. For, in general, to execute the threat will not be something A would want to do, just of itself" (italics added). This element, that compels the agent to carry out threats as well as promises, distinguishes a cooperative game for Nash; its absence distinguishes a noncooperative game. Schelling's book is firmly noncooperative in its approach, but more consciously so than much previous work, and does not simply take the noncooperative approach as given but argues for it (for example, SC, pp. 23–5, 115–18, 123–50). He also points out other implications that seem to have been overlooked before: signaling theory (SC, p. 24) deserves mention in particular.

Coordination games and coordination problems are also a major concern in the book. Coordination problems arise in games with two or more Nash equilibria. Luce and Raiffa had given examples including the famous Battle of the Sexes Game (GD, pp. 90–94; SC, 1960, p. 286 fn.) The game is shown as Table 3.3. Clearly the Nash equilibria are at the upper left and the lower right. The interest of both is in avoiding confusion that might leave them in i, II or ii, I. "What the players need is some signal to coordinate strategies; if they cannot find it in the mathematical configuration of the payoffs, they can look for it anywhere else" (SC, p. 294).

There is also a mixed-strategy Nash equilibrium, but although symmetrical, it is inferior as it imposes a 50 percent chance that both lose 1. Luce and Raiffa note, however, that if the two agents can communicate, they can arrive at a correlated strategy solution, flipping a coin and

Table 3.3   Game 3.2: the Battle of the Sexes

Payoff order: A, B		B	
		I	II
A	i	2,1	−1,−1
	ii	−1,−1	1,2

assigning a 50 percent probability to i, I and to ii, II, and zero to i, II and ii, I. Schelling's question is, however, whether they cannot coordinate their strategies even without communication, although perhaps sacrificing symmetry. A major theme of the book is that people are actually quite good at doing this. This is a focal-point solution with "some characteristic that distinguishes it from the surrounding alternatives" (SC, p. 111). The most famous example Schelling gives is the example (from his classroom experiments) of two Yale students who have to rendezvous in New York but have not agreed on the place. Most chose the information booth at Grand Central Station (SC, p. 54 fn).

Schelling's contributions cannot be contained within game theory. Schelling's thinking was also a reflection of his experience in international relations and bargaining, and the concern for practical applications in these areas recur throughout the book. The focal point idea pre-dated game theory. According to Schelling's account,[11] it arose in a student cross-country road trip about 1940 when the travelers were briefly separated. They began to think through how separated travelers might get together in a big American city in general, at that time, and decided that they could go to the general delivery window of the main post office – at 12 noon, of course.

Schelling's thinking was also a reflection of his experience in international relations and bargaining, and the concern for practical applications in these areas recur throughout the book. Even more than Luce and Raiffa, Schelling's book is mathematically informal, and he expresses some doubt that mathematical analyses are always useful rather than confusing (SC, pp. 10, 113–15, 164). Conversely, some game theorists who value mathematics highly dismiss Schelling's work as unimportant. Its influence, honored in 2005 by the Nobel Memorial Prize (Royal Swedish Academy of Sciences), cannot be denied.

3.4   TWO THEORIES, COOPERATIVE AND NONCOOPERATIVE

In 1964, Nutter proposed a noncooperative analysis of duopoly price competition in the Bertrand-Edgeworth tradition. Drawing on the growing interest in the Prisoner's Dilemma, Nutter argues that even in a duopoly, price competition is a dominant strategy equilibrium. This established a link between the Prisoner's Dilemma example and price competition that was to influence thinking in economics and industrial organization for decades to come.

In 1962 Shubik proposed that the Shapley value could be used for cost accounting in a case of shared joint costs, the first of a small stream of applications of cooperative game theory to cost assignment. This will be discussed in Chapter 8 at Section 8.2.3.

In 1963, Thrall and Lucas proposed a generalization of the game in characteristic function form, the partition function form. This form assigns a value to each coalition in a way that depends on the other coalitions that are formed, but in such a way that the value of a coalition can be different depending on the other coalitions that form. This innovation had little impact on cooperative game theory.[12] Thrall and Lucas did not present it as an alternative that could resolve the questions that had been raised about the characteristic function. Instead they followed von Neumann and Morgenstern in resolving the partition function to a characteristic function by using the assurance value. Making their theory a direct generalization of that of von Neumann and Morgenstern was a reasonable research strategy in 1963, but reintroduced the very points that had been raised against that theory. Moreover, Thrall and Lucas suggest no method by which the value of a coalition imbedded in a partition could be assigned, for example, from a representation of the game in normal form. Perhaps for these reasons, there were only a handful of extensions and applications of their theory before the 1990s. All of this justifies the neglect of partition functions by the authors of textbooks and many other cooperative game theorists, but the partition function regained some attention in the 1990s and since in game-theoretic studies of externalities.

Developments in cooperative game theory during the 1960s were important at the time and remain important for our purposes. In Aumann (CGT4, pp. 287–324), already referenced, and in several others Aumann, Davis, Maschler, and Schmeidler developed a number of new solution concepts for cooperative games: bargaining sets (Aumann and Maschler, 1964), the kernel (Davis and Maschler, 1965), and the nucleolus (Schmeidler, 1969). On Schmeidler's nucleolus and a generalization, see McCain (2013, Chapter 1.4–5 and Chapter 6). This generalization will be important in Part II.

In a number of papers Shubik, often in collaboration with Shapley, developed the theory of Edgeworth market games that he had described in the 1959 paper referenced above. Debreu and Scarf (1963) and Scarf (1967) also made important contributions, clarifying the relation of the core of a market game in characteristic function form to competitive equilibrium. These applications will be discussed in Chapter 8.

Advances in the Theory of Games (Dresher et al., 1964) contained several papers to which reference has already been made. One other that deserves mention at this point is Selten's (ibid., pp. 577–626) exploration of cooperative solutions for games in extensive form. This paper includes a discussion of the principle of backward induction (Chapter 6, section 1, below) although Selten finds it inconsistent with other, more imperative properties for cooperative solutions (Dresher et al., 1964, pp. 582, 596). He also points out – in a note added in proof – that to be complete, cooperative game theory needs the assumption that all agents can commit themselves to particular (contingent) strategies. This is a response to Schelling (1960) and is by contrast with noncooperative games in Schelling's treatment.

Experimental work on noncooperative game theory had been undertaken in the 1950s and before, some of which has been noted. Rapoport and Chammah (1965) reported a very large study of experiments centered on the *Prisoner's Dilemma*, the title of the book, a study that was to be influential. Their conception of rationality was relatively open (p. 13), and their experimental protocols required a long (but finite) series of repeated plays by the same experimental subjects, who were students. While the theory of perfect equilibrium in repeated play had not been developed, they recognized the argument that collusive agreements would "unravel" back from the last to the first repetition (ibid., p. 29). Accordingly, their observations that the noncooperative strategies were not played in any large majority of trials was seen as being inconsistent with the noncooperative equilibrium theory. In one provocative finding they discovered that female subjects cooperated less than males (ibid., p. 191). In a discussion of further experiments that might be tried, they speculated about the role a tit-for-tat strategy rule might play (ibid., p. 207). On the whole, experimental studies of the period similarly indicated that noncooperative game theory was not a strong predictor of empirical results.

In the late 1960s, Harsanyi (1967–68) introduced Bayesian reasoning into game theory in a series of three papers in *Management Science*. In 1972, Aumann and Maschler discussed some examples that raise doubts about the extensive use of behavior strategies in place of contingent strategies that had already become common. In 1972, the *International Journal of Game Theory* was founded. In the same year biologist Maynard Smith (1972) introduced the concept of evolutionarily stable strategies.

In 1973 and 1975, respectively, Gibbard and Satterthwaite published highly influential papers on the manipulation of voting schemes, in which they relied on noncooperative game theory. These will be discussed in detail in Chapter 7, at Section 7.3.2. A large literature of studies both of voting systems and of implementation of cooperative and normative objectives in terms of noncooperative equilibria has arisen from these contributions.

The 1970s were a particularly productive period for Robert Aumann, whose contributions in this period bear comparison with those of Nash around 1950. In 1973, he pointed up some difficulties with the theory of monopoly in cooperative games. In 1974, Aumann and Dreze extended and consolidated the analysis of cooperative solutions for games in coalition function form with arbitrary coalition structures (that is, partitions into distinct coalitions). This paper is the source of most of the subsequent literature on coalition structures (Greenberg, 1994) but much of the subsequent literature on coalition structures does not follow Aumann and Dreze in allowing for non-superadditive games.

In 1974 also, Aumann addressed subjective probabilities and correlated strategies. The theme of this paper is that it makes a difference if different players have different subjective estimates of probabilities of events, in the spirit of the saying "that's what makes a horse race." Aumann finds that even simple examples in noncooperative game theory must be modified to allow for these possibilities. In 1976, however, he was to argue that if two individuals have "common knowledge" of any event, which must include common prior probabilities of the event, then their posterior probabilities could not disagree – that is, "agreeing to disagree" could make no sense.

Nevertheless, Aumann's 1974 paper has been influential in another way. This paper is sometimes credited with originating correlated strategies in noncooperative games, but Aumann makes no such claim, writing (p. 70) "it has been in the folklore of game theory for years. I believe the first to notice this phenomenon (at least in print) were Harsanyi and Selten (1972)." As we have seen, the phenomenon was in fact reported in Luce and Raiffa (1957), a book which is the source of a great deal of game theory folklore. However, Aumann extends the concept with an example in which correlated equilibria can support a noncooperative equilibrium that dominates any linear (probability) combination of Nash equilibria, and this insight is the source of a stream of subsequent work on correlated equilibria. This will be discussed in detail in Chapter 5.

3.5 THE TURN TOWARDS NONCOOPERATIVE GAME THEORY

In 1975, Selten (CGT, pp. 317–54) re-examined the concept of perfect equilibrium that he had introduced in German in 1965. This paper focuses on a refinement of Nash equilibrium called the trembling hand equilibrium. Selten had introduced subgame perfect equilibrium in his paper in German ten years before, but the 1975 paper brought it to the English language audience. This paper is regarded as the beginning of the literature on "refinements" of Nash equilibrium. Selten's model will be discussed in Chapter 6 at Section 1.

In the late 1970s, work by Myerson and Maskin established mechanism design theory, following a program proposed by Hurwicz (1973). This work was honored by a Nobel Memorial Prize in 2007. See especially Maskin (1999) and Myerson (1979, 1986). The objective of mechanism design is to design a game so that its *noncooperative* equilibria correspond to the *cooperative* or other normative outcome that is desired. This seems of particular interest for public policy. Lloyd Shapley and Alvin Roth were honored by the 2012 Nobel for another sort of mechanism design, an algorithm that gives rise to pairs that are in the core of a cooperative game, and that has been implemented in a number of real clearing houses that place medical residents in positions and students in schools. Mechanism design theory will be discussed in Chapter 7.

The development of Selten's conception of perfect equilibrium made possible some important progress on what has become known as the "folk theorem" in game theory. The "folk theorem" is the idea that, for games such as the Prisoner's Dilemma (with very bad noncooperative results in one-off play) repeated play might lead to a cooperative outcome in some circumstances. As early as 1981, however, in a working paper of the UCLA department of economics, Fudenberg and Levine (1981, p. 19) sketched an analysis of repeated play of the Prisoner's Dilemma in terms of perfect equilibria. A few years later Fudenberg and Maskin (1986) gave the general analysis that has now become standard. A quite different but related approach to repeated play in noncooperative games emerged with Axelrod's (1981, 1984) computational studies. Coding simple rules for the selection of behavior strategies in repeated Prisoner's Dilemmas, Axelrod played the rules one against another in a tournament, and found that tit-for-tat[13] (a trigger strategy in which one plays cooperatively until the first defection by the other player, but responds with a single round of non-cooperative play) did relatively well against a wide array of challengers. As much as the folk theorem work, this study contributed to the emergence of tit-for-tat and other trigger strategies as standard tools for understanding repeated play of noncooperative games.

In 1984 Bernheim introduced the concept of rationalizable strategies; a simultaneous paper of Pearce (1984) shared the innovation. One important departure of this paper is that Bernheim allows players to condition their decision rules on conjectures about the conjectures that others may make about them. This leads, in some cases, to a much larger set of stable strategies. In his Nobel Address, Aumann (2005) was to admit conjectures as to the *rules other players might use in selecting behavior strategies* among the conditions of a choice of strategies, with a further extension of the range of possible noncooperative equilibria in repeated games.

When we combine rationalizable strategies and correlated equilibrium, the case for Nash equilibria as predictors of behavior is very much reduced. If the game is played one-off, then players are not likely to have enough information to exclude non-Nash rationalizable equilibria, and the same will be true in the first plays of a repeated game. For later plays of a repeated game, though, correlated equilibria may emerge, and these, too, may be non-Nash. In an evolutionary model, where players are randomly matched to play one-off but can learn from the experience of one another, evolutionarily stable (Nash) equilibria seem a reasonable prediction. Even here, though, boundedly rational learning might result in correlated equilibria. More generally, where the Nash equilibrium in pure strategies is unique (including, but not limited to, the family of social dilemmas) correlated strategy equilibria can be excluded; and if in addition the Nash equilibrium is subject to some stringent stability conditions (Bernheim, 1984, p. 1020) then Nash equilibria are the only rationalizable strategies. All in all, Nash equilibria can no longer be treated as "solutions" to non-cooperative games, but only as candidate solutions and as tools that may be useful in finding other (for example, correlated equilibrium) solutions.

In 1988, Harsanyi and Selten offered a framework to resolve the growing family of refinements of Nash equilibrium, suggesting a hierarchy of criteria for choosing among Nash equilibria. They rank the equilibria in terms of relative stability, so that, for example, Pareto-dominant equilibria are considered more stable than those that are not Pareto-dominant but are risk-dominant.

In 1989, the journal *Games and Economic Behavior* was founded.

3.6   BEHAVIORAL GAME THEORY

Traditional game theory proceeds from strong assumptions about human rationality to strong conclusions about the nature of equilibrium. One can ask whether either the assumptions or the conclusions are empirically valid. If we find evidence that they are not, and attempt to rebuild game theory with more "realistic" assumptions about rationality, we are entering the sphere of *behavioral game theory*. The more traditional studies based on those strong assumptions will be called *classical game theory*.

There is a long history of experimental studies in game theory, some of which have been mentioned in context. Social psychologists and others quite early provided evidence that people facing a Prisoner's Dilemma-like game do not always act as neoclassical maximizers (Lave, 1965; Rapoport and Chammah, 1965; Morehouse, 1967; Kreps et al., 1982, among many others). For an argument that game theory ought nevertheless be based on strict rationality, see Morgenstern and Schwödiauer (1976). Some scholars suggest that even the successful trials are attributable to training effects (Marwell and Ames, 1981; Carter and Irons, 1991). But game theoretic equilibria also gain some experimental support, especially in their evolutionary interpretation (for example, Cooper et al., 1990; Van Huyck et al., 1990). Mailath (1998) surveys evolutionary game theory, to determine the extent to which it may support the predictions of Nash equilibrium in particular, indicating that "Evolutionary game theory has provided a qualified answer . . . In a range of settings, agents do (eventually) play Nash" (p. 1348). However, he also indicates the limits of this range.

Some of the early experiments on the Prisoner's Dilemma were interpreted as evidence that altruism is an element in human behavior. Unfortunately, altruism is not always well-defined. Altruism was inferred, however, from a tendency to choose the cooperative strategy even when it is not an equilibrium strategy, for example, in Prisoner's Dilemma games. More recent studies have often focused instead on reciprocity. Berg et al. (1995, p. 139) say their " . . . results suggest that both positive and negative forms of reciprocity exist and must be taken into account . . . [and] provide strong support for current research efforts to . . . integrate reciprocity into standard game theory." Positive reciprocity means that players respond to generous behavior generously, even at a sacrifice to themselves; negative reciprocity means that they retaliate against aggressors even when it makes them worse off to do so.

One game that has been extensively studied in the experimental literature is the "Ultimatum Game" (for example, Henrich et al., 2005). The Ultimatum Game is a two-person game along the following lines: the two agents may be able to share a fixed amount, such as $100. The first agent, the proposer, suggests a payment to go to the second agent, the responder. If the responder accepts the payment, he receives it, and the balance is paid to the proposer. However, if the responder rejects the payment, neither agent gets anything. The noncooperative equilibrium is one in which the proposer makes the smallest possible positive offer and the responder accepts it. However, experimental evidence disagrees with this prediction. If the proposer makes a very small offer, the responder is sometimes observed to reject the proposal despite sacrificing the small positive payment. Moreover, offers are often more than the minimum needed to avoid a rejection, and 50–50 offers are fairly common.[14] This and other experimental games have given rise to their distinct specialized literatures.

In all, the experimental evidence does not support many of the predictions of noncooperative game theory. This is not to say that it supports any of the cooperative solutions in any systematic way, either. We must suppose that real human beings are both more complex and less accurate calculating machines than classical game theory supposes. Solution concepts based on strict rationality can define hypotheses as to attractors and stable points in dynamic models with boundedly rational learning. On that score the more recent experimental evidence is not merely negative. Models based on strict rationality also define the base from which deviations from rationality are predicted.

3.7 TOWARD UNITARY GAME THEORY?

In the 1990s and 2000s, advances continued to be made in the topics of noncooperative and cooperative game theory that had come to be traditional, and some research pursued new directions that will be useful for this book. Returning to the long-neglected topic of coalitions in noncooperative games, Bernheim et al. (1987) proposed a property of coalition-proofness as a refinement of Nash equilibrium. The period was, of course, dominated by the Nobel Memorial Prizes of 1994, 2005, 2007, and 2012, which kept the traditional topics in view. Nevertheless, in his inaugural presidential address to the Game Theory Society, Aumann (2003) expressed regret at the division of game theory between cooperative and noncooperative branches, saying that there ought to be one theory of interactive rationality. Certainly the division weakens game theory by its ambiguity, but more than that, common experience tells us that both noncooperative and cooperative actions are parts of our experience. On the one hand, in Maskin's (2004) words, "We live our lives in coalitions." On the other hand, tragedies of the commons and other social dilemmas, involuntary unemployment and price competition are no less parts of our experience.

Some works have addressed this division. In 1990, Greenberg proposed a "theory of social situations" as an alternative both to cooperative and noncooperative game theory. Some progress was made on the incorporation of externalities in games in partition function form. Zhao (1992) proposed a theory of "hybrid solutions." Chwe (1994) addressed problems that arise for a core-like solution in games in partition function form, and Ray and Vohra (1999) proposed a theory to explain the determinants of the coalition structure.

Brandenburger and Stuart (2007, for example) proposed "Biform Games," a two-stage analysis in which the first stage is a noncooperative game, but the outcome of the noncooperative game is a cooperative game that determines the payoffs to the participants. The solution to the two-stage game is by backward induction. This sort of approach actually has a long history: von Neumann and Morgenstern (1944/2004) argue along these lines in their simple majority game example (p. 222). Nash (1953) variable threat bargaining theory is a formally similar two-stage game. However, Brandenburger and Stuart, working mainly on issues of corporate strategy, show how this approach permits a reconciliation of cooperation within coalitions with such noncooperative phenomena as inefficient externalities. Applying a different bargaining theory, McCain (2013, Chapter 7) extends these examples, and this discussion will play an important part in Part II of this book.

3.8 BRIEF SUMMARY

In 45 years from 1944 to 1989, game theory became a cross-disciplinary study of great importance for the mathematical social sciences. It also became a compound field – not one study of interdependent decisions, but largely separate studies of noncooperative and cooperative game theory.

What game theory offers is a kit of tools applicable to decision problems in which the consequences of one decision may depend on the decisions of others, previous decisions creating the conditions for current decisions, simultaneous and subsequent decisions, with or without mutual knowledge, with or without some degree of honest mutual commitment to a common strategy. If we choose our tools to fit the job and disregard the dogmas and dichotomies of cooperative and noncooperative, superadditive valuations and rationality, we will find that the tools contribute to the solution of problems of real-world public policy.

NOTES

1. Where two dates are given, the first is the original date of publication, and the second the date of the edition or translation used.
2. Kyu Uck Lee, personal communication by e-mail, June 22, 2007.
3. "A Chronology of Game Theory," by Paul Walker, Economics Department, University of Canterbury, Canterbury, New Zealand, http://www.econ.canterbury. ac.nz/personal_pages/paul_walker/gt/hist.htm, accessed 9/9/2014, last modified Sept. 2012.
4. In what follows page citations indicated by TGEB will refer to von Neumann and Morgenstern (2004).
5. This paper seems to have been available only in German prior to the publication of the translation in Tucker and Luce (1959).
6. In what follows citations to CGT will refer to Kuhn, *Classics in Game Theory*, 1997.
7. In what follows page references indicated by CTG4 will refer to volume 4 of *Contributions to the Theory of Games*, edited by Tucker and Luce.
8. In what follows references denoted GD will refer to Luce and Raiffa (1957).
9. Von Neumann and Morgenstern had alluded to this at TGEB, p. 224.
10. In what follows page references to SC refer to Schelling, *Strategy of Conflict*, 1960.
11. This paragraph relies on an address given by Schelling at Trinity University, San Antonio, Texas on April 18, 2007, and is from memory.
12. One evidence of this is that it is not mentioned in some recent advanced texts with coverage of cooperative game theory. See, for example, Peleg and Sudhölter, *Introduction to*

the Theory of Cooperative Games, 2003, and Forgo et al., *Introduction to the Theory of Games*, 1999.

13. According to the Oxford English Dictionary, the phrase tit-for-tat is traceable to the 16th-century phrase "tip for tap," meaning, roughly, push for shove.

14. For example, Guth et al. (1982), Henrich et al. (2005), and note also Roth et al. (1991), Roth and Erev (1995), Stanley and Tran (1998), Oosterbeek et al. (2004), Andreoni and Blanchard (2006).

4. Nash equilibrium and public policy

The best-known ideas in game theory are within noncooperative game theory, and probably the single best-known example in game theory is the Prisoner's Dilemma, a noncooperative example. This example shows how interactive self-interested decisions may lead to results that are less favorable to all participants than some other outcome would be. The Prisoner's Dilemma example can be generalized to a class of noncooperative normal form games known as "social dilemmas" (Dawes, 1980) that share similar broad qualities. From the pragmatic point of view, noncooperative game theory provides powerful tools for the identification and specification of problems, as the social dilemmas exemplify. On the whole, moreover, noncooperative game theory is a relatively settled, mature study. Social dilemmas are a class of Nash equilibrium models, and Nash equilibria are well understood and the foundation of most applications of noncooperative game theory. However, there are some unsettled issues and some other proposed approaches to the solution of noncooperative games. This chapter will review a number of Nash equilibrium models with a view to their applicability to public policy studies.

4.1 SOCIAL DILEMMAS

While the Prisoner's Dilemma is the best-known example in game theory, it is also one of the simplest, and its simplicity does place some limits on its application.

4.1.1 Symmetrical Dilemmas

The Prisoner's Dilemma begins with a story of interrogation. For this discussion, we may instead recall the Water Game from Chapter 2, where it is shown in normal form as Table 2.1.

Eastland knows that it cannot influence Westria's strategy choice, and conversely. Instead, each one chooses his best response to the strategy choice made by the other. This defines *Nash equilibrium*. Moreover, in this case, the best response is "Divert" regardless of the other agent's strategy choice. That means "Divert" is a *dominant strategy*: by definition, if a strategy is the best response to any strategy choice made by the other agent or agents, it is a *dominant strategy*. Thus, when both agents choose the dominant strategy "Divert," we have a *dominant strategy equilibrium*, which is a particularly simple instance of a Nash equilibrium. A dominant strategy equilibrium can be defined as a Nash equilibrium in which each agent has a dominant strategy.

Nevertheless, if both agents were to choose "Don't" in Game 2.1, both would be better off, with net payoffs of 0 rather than -1. We may borrow terminology from welfare economics and say that the strategy pair "Don't, Don't" *Pareto-dominates* the pair "Divert, Divert." A strategy vector Σ_1 Pareto-dominates strategy vector Σ_2 if no agent is worse off with Σ_1 than with Σ_2 and at least one agent is better off with Σ_1 than with Σ_2. Together, these observations define Game 2.1 as a *social dilemma* (Dawes, 1980). Generally, a social dilemma is a game in which (1) there is a dominant strategy equilibrium indicated by strategies Σ_2, and (2) there is vector of strategies Σ_1, such that each component of Σ_1 differs from the corresponding component of Σ_2 and Σ_1 Pareto-dominates Σ_2. Social dilemmas are usually also treated as being symmetrical (so that interchanging any two agents would leave the payoff table unchanged).

A social dilemma model such as Game 2.1 predicts that, in the absence of some public intervention, the dominant strategy equilibrium, Σ_2, will occur. Since it is Pareto-dominated by a different set of decisions, Σ_1, this outcome is *inefficient*. Decisions Σ_1 are said to constitute a *cooperative solution* and, in a symmetrical game such as this, the payoffs correspond to the Shapley value and nucleolus in particular. (These will be discussed in more detail in Chapter 8. Note also McCain 2013, Chapters 1, 5, 6.) Public policies may then be advocated that move individual decisions toward the efficient set Σ_1. In this way, social dilemmas capture the principles that seem to underlie a number of major problems of modern societies and public policy, but they may not be very good descriptions of the real world. The symmetry that is usually assumed in social dilemma models is one example. In most real-world applications, there is likely to be some lack of symmetry among the agents. Since the problem (inefficiency) arises in symmetrical models, however, we can be assured that it does not arise because of the lack of symmetry in the real world; thus asymmetry is a complication but not an underlying cause of the problem. This is a valuable point that might be missed if the simplified, symmetrical model were not considered. All the same, for some practical applications, it may be necessary to reintroduce some asymmetry in a model with heterogeneous agents.

Social dilemmas can be generalized to a large number of players following Schelling (1978) and Moulin (1982 p. 92 *et seq.*). Think of a large number of people living in the watershed of a lake, each of whom may act

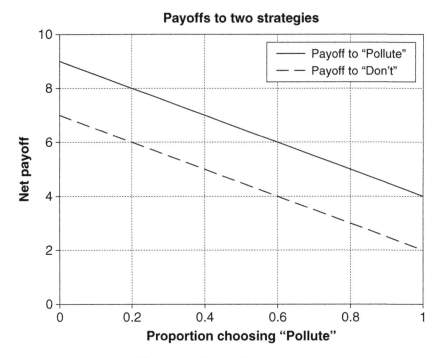

Figure 4.1   A social dilemma with large N

so as to pollute the lake or, at some cost, refrain from pollution. Suppose there are N agents, N very large, each of whom must make the same absolute choice of strategies "Don't" or "Pollute." The overall amount of pollution will depend on the proportion of the population that choose "pollute;" so that the payoffs to both strategies will depend on the same proportion. Borrowing terminology from the theory of differential games, we can describe the proportion of agents who choose "Pollute" as a *state variable* for the game. In this usage, a state variable is a variable that is sufficient to determine the payoffs of the different strategies without any other information (such as information on the specific strategy choices of individual agents, for example).

This model is illustrated by Figure 4.1. We see that the payoff to "Pollute" lies above the payoff to "Don't," regardless of the proportion of the group who choose "pollute" as their strategy. The diagram illustrates visually that this is an N-person social dilemma. If any group of players chooses "Don't," they are not choosing their best response to the strategies chosen by the others. The dominant strategy equilibrium corresponds to the rightward extreme of the diagram, the case in which every agent chooses "Pollute."

The N-person social dilemma model can also be interpreted to be consistent with imperfectly rational behavior, if it is interpreted in an evolutionary sense (see, for example, Aumann 1997). Suppose that individuals usually act with inertia, simply choosing the same strategy over and over, but from time to time, at random, they experiment with reversing their strategies. If the reversal leads to an increase in the net payoff they persist, and if not they return to the previous strategy. This trial-and-error learning process is one of random *variation* and directed *selection* of strategies, a simple evolutionary process. Biologists, having borrowed the concept of Nash equilibrium from game theory, define an evolutionarily stable strategy (ESS) as a Nash equilibrium that is stable under an evolutionary dynamics. The dominant strategy equilibrium for this model is an ESS. Thus the conclusion does not depend on the assumption of perfect rationality.[1]

An appropriately generalized social dilemma model can account for many instances in which inefficiencies persist in the presence of human decisions that successfully seek self-regarding benefits, whether through perfect rationality or through trial-and-error learning. As this example and the Water Game suggest, environmental economics is largely built of social dilemma models. The production of a public good is another instance of a social dilemma. All in all, social dilemma models are powerful diagnostic and explanatory tools for problems of social inefficiency.

4.1.2 The Special Case of Price Competition

This may seem a bit strange to a person whose knowledge of game theory is derived from a textbook of microeconomic principles. In microeconomics price competition leads to efficiency, and price competition is noncooperative behavior; cooperative behavior (collusion) is inefficient and is a danger to be avoided. But price competition is very much a special case, which mixes cooperative and noncooperative elements right from the start.

A discussion of price competition begins by assuming that a relatively small number (but greater than one) of coalitions called "business firms" make offers to sell some product. Their strategies may be the offer prices (the case known as Bertrand-Edgeworth competition) or the quantities offered (Cournot competition). The buyers are not usually modeled as individual agents but treated as an undifferentiated mass of demand. Implicitly, however, the buyers are treated as acting as singleton coalitions, and the strategy of each buyer is the quantity she or he chooses to buy. If, then, the sellers' behavior is characterized by noncooperative Bertrand-Edgeworth competition and the singleton buyers act noncooperatively, and if there are no externalities, the outcome is efficient. (If seller competition is Cournot, then the outcome approaches efficiency as the number of sellers increases and their sizes approach equality.) On the other hand, if the sellers form a coalition and act cooperatively among themselves, restricting output and raising the price, the outcome is inefficient.

Nevertheless, price competition is not really a noncooperative game, since the business firms are coalitions based on legally enforceable agreements; and because exchange is itself a cooperative activity. To use Brandenburger and Nalebuff's (1997) phrase, this is "co-opetition." Price competition can best be understood as a hybrid solution (Zhao, 1992) for a game with a coalition structure (Aumann and Dreze, 1974). In principle, the grand coalition of all producers and consumers could obtain the efficient outcome, but free-market economists believe (plausibly) that this degree of centralization is nonfeasible. If so, then the hybrid solution of price competition may be our best hope for efficient markets.

The key lessons to be drawn are that there is no real contradiction between the recommendation of price competition in microeconomics and the recommendation of cooperative solutions in game theory in general, and that the results of efficient market theory cannot be generalized to other cases without strong independent justification for doing so. The neo-liberal attempt, in the later twentieth century, to organize all social action on the basis of price competition may ultimately be seen as no less utopian than the attempt, earlier in the twentieth century, to organize social action on the basis of a communistic grand coalition.

4.1.3 Other Dilemmas, Nash Equilibria and Public Policy

Social dilemmas seem to be at the basis of many environmental and other problems, but there is a tendency to overuse the social dilemma model. This is confusing because there are many models based on the Nash equilibrium that are not social dilemmas (because some players do not have dominant strategies or because asymmetry is important, for example) but in which the Nash or other noncooperative equilibrium is inefficient.

Let us consider another example to do with riparian rights and water supplies. For this example, the two players are two municipalities in the same river valley. The Town of Upstream is, as the name suggests, upstream from the Boro of Downstream. Upstream can obtain its water supply by damming Modest River at a point well above Downstream, but to do so would impair Downstream's ability to meet its own water needs from the Modest. Geography is such that Downstream could build either one or two dams (or none). Although two dams would be costly, together they could

Table 4.1   Game 4.2: water works

Payoff order: Downstream, Upstream		Upstream	
		One dam	No dam
Downstream	One dam	3,3	5,1
	Two dams	2,3	4,4
	No dam	1,5	1,1

supply both towns, provided the Upstream dam is not built. Game 4.2 is a payoff table based on those ideas, with the benefits from water provision rated on a scale of 1 to 5 for each town (see Table 4.1).

For Game 4.2, the Nash equilibrium is for each town to build one dam. We may observe that for the Boro of Downstream, building one dam is a dominant strategy: regardless of the strategy chosen by Upstream, building one dam gives higher payoffs than any other strategy. On the other hand, Upstream does not have a dominant strategy: either one strategy or the other may pay best, depending on the strategy chosen by Downstream. However, Upstream can make some reasonable judgments about the strategy Downstream will choose. Upstream knows (after all) that "two dams" and "no dams" are dominated strategies, and also that Downstream is a rational decision maker. Thus, Upstream can draw the reasonable conclusion that Downstream will never choose those dominated strategies. They can be left out of consideration, since they will not affect the game. Once they are dropped out, we have an equivalent game in which Downstream has only undominated strategies, and the only one is "one dam." In that smaller game, the strategy of building one dam is dominant for Upstream. We then eliminate the dominated strategy for Upstream, and are left with only one strategy combination: "one dam," "one dam." This is an instance of the *iterated elimination of dominated strategies*. A strategy is said to be *strictly dominated* for Player i if it yields a payoff to i that is greater than the payoff to any other strategy, regardless of the strategies that other players choose, and is said to be *weakly dominated* if it yields a payoff no less than that from any other strategy i may choose. When we eliminate the strictly dominated strategies for a player, creating a reduced game, and repeat that so long as it is possible, we are applying *iterative elimination of (strictly) dominated strategies*, and if that procedure yields a unique strategy for each player, those strategies correspond to a unique Nash equilibrium.

As in a social dilemma, payoffs to both players for the Nash equilibrium are dominated by the payoffs that result if Downstream builds two dams and supplies Upstream from its surplus water. Nevertheless this is not a social dilemma. As we have seen, Upstream does not have a dominant strategy. For Upstream the best response is to build a dam if Downstream builds less than two. And this makes a difference. In Game 4.2, we have assumed that both decisions are made simultaneously, with neither decision maker aware of the decision the other has made or will make. Suppose instead that one of the decision makers can commit himself by making the first move. In that case the second mover can choose contingent strategies, using his knowledge of the first-mover's decision. In the social dilemma this makes no difference, since each decision maker's best strategy is the same no matter what the first mover may do. In Game 4.2, by contrast, if Downstream is first mover we will see a different outcome. Downstream can anticipate that, if they build no dam or one dam, Upstream will build its dam, leaving the Downstream with a payoff of 3 at most; while if Downstream builds two dams and supplies both towns, Upstream will not build its dam, giving Downstream a payoff of 4. For Downstream as first mover, the strategy of building two dams becomes a best response to the contingent strategy "if one or fewer Downstream dams, then build, else do not" on the part of Upstream.

It is important, then, not to generalize the social dilemma too hastily. Not all social problems are social dilemmas, and that may make a difference in the attainable solutions.

4.2   RANDOMIZATION OF STRATEGIES

In the examples in this chapter so far, each decision maker has to choose among a finite number of "strategies," where each strategy is a description of a course of action, with or without a description of the contingencies in which that action will be taken; that is, either behavior strategies or contingent strategies. These "strategies" are often called "pure strategies," but in this book the word "strategies" will mean pure strategies unless it is indicated otherwise. Indeed, thus far in the chapter, the number of strategies is quite small and decisions are simultaneous so that contingencies will not need to be specified. In each case so far, there is at least one Nash equilibrium in pure strategies. However, even in games with small numbers of strategies and simultaneous decisions, there may not always be a Nash equilibrium if we limit ourselves to pure strategies.

We can illustrate that with an example that was historically important, in that it captured Oskar Morgenstern's attention and led to his collaboration with von Neumann. The example is from the Sherlock Holmes story, "The Final Problem." Holmes is attempting to escape from Moriarty and from England to the continent, and Moriarty is attempting to capture

Table 4.2   Game 4.3: "The Final Problem" in strategic normal form

Payoff order: Moriarty, Sherlock		Sherlock	
		Canterbury	Dover
Moriarty	Canterbury	1, −1	−1, 1
	Dover	−1, 1	1, −1

and murder Sherlock. Sherlock is on the train to Dover and must decide whether to stay on the train and cross to Europe via Dover or to get off at Canterbury and go to Europe by a different route. Moriarty must decide whether to continue to Dover and try to intercept Sherlock on the coast or to leave the train at Canterbury in the hope that he will find Sherlock there. If Moriarty can choose the same stopping point as Sherlock, then Sherlock is caught; otherwise Sherlock can escape to the continent. Assigning a payoff to Moriarty of 1 if he catches Sherlock and −1 otherwise, and to Sherlock of 1 if he escapes and −1 otherwise, we have the game in normal form shown as Game 4.3 in Table 4.2.

This game has no Nash equilibria in the pure strategies "Canterbury" and "Dover." If Moriarty knows that Sherlock will detrain at Dover, then he also will detrain at Dover, but in that case Sherlock's best response is Canterbury, though if Sherlock gets off at Canterbury, Moriarty's best response is to do that too – and so on! In short, there are no two strategies each of which is a best response to the other: no Nash equilibrium in pure strategies. More generally, if either Sherlock or Moriarty acts predictably, he is likely to lose out, as the other person can use that predictability to defeat him. Indeed, the best that either one can do is to choose between the two strategies at random, assigning a probability of one-half to each strategy. If (for example) Sherlock deviates from that 50–50 probability, choosing Dover with a probability greater than one half, he loses out on the average, since Moriarty can get off at Dover with a better than even chance of catching Sherlock.

When a player chooses between two strategies at random, according to probabilities that prevent the other player from exploiting his predictability, this is called a *mixed strategy*. There are infinitely many mixed strategies in a nontrivial game, and Nash showed that every game in normal form has at least one Nash equilibrium, which may be a mixed strategy equilibrium. Nash equilibria in mixed strategies can be computed by linear programming methods in general, or, in simple cases, by the algebra of simultaneous equations. Details of computation will be beyond the scope of this chapter.

Table 4.3   Game 4.4: Terrorist vs. Defender

Payoff order: Defender, Terrorist		Terrorist	
		Target 1	Target 2
Defender	Target 1	3,0	0,3
	Target 2	0,3	3,0
	Both	2,1	2,1

In situations of conflict, it is often in the interest of each party to act unpredictably. The Normandy Invasion, Game 2.6, provides another good example. It illustrates an important point: it is not necessary literally to flip a coin to decide between two strategies. The probabilities that matter are subjective probabilities. As a player in the game, my objective is to manipulate the subjective probability estimates my enemy assigns to my own actions, so as to prevent my enemy from exploiting the predictability of my decisions. The web of costly deception and strategic maneuvers surrounding the Normandy invasion, as described in Brown (1975), *A Bodyguard of Lies*, illustrates this.

Let us consider one further example in that vein. The players in Game 4.4 will be a terrorist and a defender.[2] The terrorist has the capacity to attack target 1 or target 2, but not both and no other target. The defender can prevent an attack on target 1 or target 2 by appropriate defensive measures, or, at somewhat greater cost, can protect both. Complete success for either contestant is recorded as a payoff of 3. If the defender defends both targets, the defender's payoff is reduced to 2, because of the cost of doing so, and this partial success for the terrorist is recorded as a payoff of 1. Notice that the sum of the payoffs is 3 in every cell of the table (Table 4.3). Although this is not a zero-sum game, it is a constant-sum game, and just as with a zero-sum game, this means the objectives of the players are totally opposed.

Suppose, then, that the terrorist chooses a mixed strategy, assigning probability ½ to each target. Then the defender's expected value for defending a single target is.

$$\frac{1}{2} \cdot 3 + \frac{1}{2} \cdot 0 = 1\frac{1}{2}.$$

Then the defender has no reason to defend one target in preference to the other: to that extent the mixed strategy has done its job. By defending both targets the defender can have an expected value of 2. The defender's equilibrium strategy is to defend both targets. By randomizing their strategy the terrorists have forced the defender to the expense of defending both targets, gaining a minor victory by doing so.

Mixed strategy equilibria in peacetime public policy problems may be uncommon, but the possibility should not be overlooked. Whenever we see decision makers creating doubt about their strategies, we will need to explore the possibility of a mixed strategy equilibrium. In relations between local governments and large businesses or major athletic teams, for example, it may well be that the uncertainty the firms and teams create about their locational decisions represent a mixed strategy.

4.3   COORDINATION AND ANTICOORDINATION GAMES

We may think of the game as a mathematical problem and Nash equilibrium as a solution. In that perspective the Nash equilibrium in pure strategies has two kinds of shortcomings. First, as we have seen, some games may not have Nash equilibria in pure strategies. When we allow mixed as well as pure strategies, however, that problem disappears: as a matter of mathematical fact, every game in strategic normal form has at least one Nash equilibrium. (As a practical matter, though, randomized strategies may have to be excluded in some special case applications, and in that case the difficulty returns.) The other shortcoming is that there may be more than one Nash equilibrium. The right number of solutions, from a mathematical point of view, is exactly one. Indeed some critics of game theory have made that shortcoming the basis of a claim that the rational action model of human behavior (as expressed in Nash equilibrium) is simply a failure. But that is a hasty conclusion. On the one hand, the plurality of solutions may reflect the conditions of the real world, rather than a failure of mathematics. In that case we would not want a solution that, however perfect mathematically, assumes away the facts of the real world. On the other hand, we may treat the multiplicity of solutions as a problem to be solved, not by the theorist but by the "players in the game," and inquire how in fact people have contrived to solve it. This proves to be a rich field of inquiry, one we will undertake in the next chapter.

A class of games that illustrates both these points is the so-called coordination games. Examples can be drawn from highway traffic. As usual, we may begin with a simple two-person game in which the two agents must each choose between two strategies. The two agents will be motorists approaching one another on a road, and the strategies they must choose between are "drive on the left" and "drive on the right." If they make the same choice, then they pass one another safely. If they make opposite

Table 4.4   Game 4.5: the fender-bender game

Payoff order: First, Second		Second car	
		Left	Right
First car	Left	1,1	−1,−1
	Right	−1,−1	1,1

choices, then they both lose in the resulting fender-bender. The game is shown in strategic normal form as Game 4.5 (Table 4.4). The payoff numbers are arbitrary: 1 for a safe passage and −1 for a collision.

In this game there are two Nash equilibria in pure strategies: "left," "left" and "right," "right." In addition there is a third Nash equilibrium, a mixed strategy equilibrium in which each motorist chooses randomly with probability one-half for each strategy, and the expected value of the payoff for each motorist is zero. Clearly, the pure strategy equilibria are superior to the mixed strategy equilibrium or to any non-equilibrium state. (They are superior in the sense that the pure strategy equilibria are Pareto-dominant over the others.) Now suppose that each motorist knows *nothing* that has not already been given as part of the game. Then each may reason as follows: "Using the principle of insufficient reason, I must assign equal probabilities to the other person's strategy choice. Therefore I have nothing to lose by also choosing my strategy at random." This points to the inferior mixed strategy equilibrium as the most likely one, and that is indeed an ugly dilemma (although not a "social dilemma").

When we assumed that each motorist knows nothing that has not already been given as part of the game, we assumed a great deal. In particular we assumed that the drivers do not know whether they are driving in North America, England, India, or Europe. In each of those countries it is *customary* to drive on one side or the other, and, knowing the custom, each agent can make a rational judgment that the other will (with very high probability) drive on the customary side. The result is that fender-benders in this situation are actually quite rare. The familiar fact that the custom can be quite different in different countries – left in England, right in the USA – reflects the fact that both are pure-strategy Nash equilibria. What seems a shortcoming from the mathematical viewpoint proves to be a principle with explanatory power in this application.

Of course, there are also legal standards in most countries *requiring* people to drive on the customary side. But the customs are mostly self-enforcing, and the function of the laws is mainly to assign responsibility when the custom fails and collisions do happen. Simple as this example is, it suggests reasons why customs, much as they may vary from country to country, can be very stable where they exist, and why the law may be most effective when it reinforces custom and may be largely futile when it goes against custom.

It will be useful as well to view this example in an evolutionary light. Suppose that we have a large population of motorists who are randomly matched to pass one another on the road in a large series of matches. Suppose also that motorists choose their strategies according to some boundedly rational rule of thumb. One possible rule of thumb is conformism: "Do what you see others doing." Another possible rule of thumb is the stick-or-switch rule: "If the strategy yields a positive payoff, stick with it; otherwise switch." Either of these rules will lead to a rapid convergence to the unanimous choice either of left or right, though we cannot predict which.

Games of this kind are called *coordination games*, since the problem faced by the two players is to coordinate their strategies and thus avoid a bad outcome. Following local custom is one instance of the kind of solution suggested by Thomas Schelling (1960) and is sometimes called a "Schelling point" or a "focal equilibrium."

Now let us consider another two-by-two game, again involving two motorists. In this case, however, the two motorists are approaching an intersection, and their strategies are to stop and let the other go through, or to go ahead. There are four possible outcomes: a fender-bender if both go, a waste of time if both stop, and two outcomes in which one is slightly delayed and the other passes without delay when they choose opposite strategies. This is shown as Game 4.6 (Table 4.5), with the payoff numbers assigned arbitrarily (as usual) to represent better and worse outcomes from the different points of view of the two motorists.

In some ways this game is very much like the coordination game: it has two Nash equilibria in pure strategies and a third, mixed strategy equilibrium, in which each driver chooses between the two strategies with probability ½. The expected value of the mixed strategy is −½, so the mixed strategy equilibrium is Pareto-dominated by the pure strategy equilibria. Once again, though, with no information but what is contained within the game, the two motorists face a problem in choosing strategies

Table 4.5 Game 4.6: the intersection game

Payoff order: First, Second		Second car	
		Go	Stop
First car	Go	−2,−2	1,0
	Stop	0,1	−1,−1

that will avoid the bad outcomes. In this case, though, they have to choose *opposite* strategies, and for this reason games of this kind are sometimes called *anticoordination* games.

In some ways, though, this is a more difficult problem. In a coordination game, coordination can be accomplished when both motorists receive the same signal, as with a custom that all motorists drive on the right. For this case, though, the two motorists would have to receive different signals, or a signal complex enough that they would interpret it in opposite ways. For the same reason, a simple evolutionary process may not lead to the pure strategy Nash equilibria in this case. Suppose that a large population of motorists are randomly matched to play the two-person game, again and again. Clearly, a rule of conformism will not lead them to make opposite decisions. The mixed strategy equilibrium proves to be evolutionarily stable, since any tendency for more than half of the motorists to choose one strategy rather than the other just makes coordination of the strategy choices less likely! Anticoordination games are a problem to which we will return in the next chapter.

4.4   RATIONALIZATION, ERRORS, AND NASH EQUILIBRIUM

One of the key ideas of game theory is that a rational person must not only try to anticipate the rational choices of others, but do so on the understanding that they also are attempting to anticipate one's own rational choice. This is captured in a very broad sense in the concept of rationalizable strategies (Bernheim, 1984; Pearce, 1984; McCain, 2014b, Chapter 4, sections 4.5–4.10, Chapter 11, sections 11.8–11.9). Let the players be a and b. Then a strategy is rationalizable for a if it is a best response to a strategy b might rationally choose. But, in turn, b will rationally choose the strategy only if it is a best response to a strategy that a might rationally choose; this in turn means that it is a best response to a strategy b might rationally choose. If this circle can be closed, then the strategy with which a began is a rationalizable strategy for a.

Strategies corresponding to a Nash equilibrium are always rationalizable since each player is choosing a best response to the other's best response. However, not all rationalizable strategies correspond to Nash equilibria. (For examples to illustrate this see McCain, 2014b, pp. 80, 255–60.) Nevertheless, the Nash equilibrium is a condition of stability. If the players in the game choose rationalizable strategies that are not Nash-equilibrial, at least one will find that she has made a mistake. If the strategies are not Nash-equilibrial, it follows that at least one has chosen a strategy that is *not* a best response to the strategies chosen by the others. By assumption the strategy was rationalizable, and so it is a best response to the strategies she expected the others to choose. Thus, her expectations can only have been mistaken.

This has two implications. First, for some decisions, there is no possibility to correct the error – the decision is made at a particular time and place and is determinate once the error is known. For such decisions the Nash equilibrium may not be an appropriate solution concept. Second, if the circumstances do permit the players to correct their errors, then non-Nash rationalizable strategies will be unstable under the learning process. This is the meaning of the phrase "Nash equilibrium is a condition of stability." This is no less true if there are two or more equilibria in pure strategies, as in coordination and anticoordination games. However, when there are two or more Nash equilibria, the beliefs of the players about one another will be different. Each set of beliefs in a Nash equilibrium is consistent, in that it leads each player to act in ways that confirm the expectations of others, and each set of consistent beliefs shares this stability under the learning process.

In coordination and anticoordination games, consistent beliefs are sufficient for efficient action and a cooperative outcome in the game. The problem is to increase the probability that beliefs will be consistent. In a coordination game, a common custom achieves this. The common belief that "in North America, it is safest to drive on the right hand side of the road" proves to be a true belief – not because it is a truth in any metaphysical sense, but because the actions that people choose, on the basis of the belief, make it a true belief. In a social dilemma, by contrast, consistent beliefs are not sufficient for efficient and cooperative action, and in fact correspond to an inefficient outcome. This is an important distinction for public policy, since in the first case, the case of coordination games, the efficient arrangement is self-enforcing, while in the second case, the social dilemma, enforcement must play the main role in achieving an efficient cooperative outcome.

4.5   COALITIONS IN NONCOOPERATIVE GAMES

In noncooperative games, there are no enforceable agreements. Nevertheless, when there are two or more Nash equilibria, coalitions may form and may make a difference. Consider Game 4.7, of conflicts among three nations (Table 4.6), with the following assumptions: Country a is the strongest of the three, and capable of projecting both land and sea power, Country b is landlocked, and thus unable to influence the balance

Table 4.6   Game 4.7: conflict among three nations

Payoff order: Country a, b, c	c			
	Forward		Defensive	
	b		b	
	Forward	Defensive	Forward	Defensive
a Forward	1,1,1	7,3,1	7,1,3	8,3,3
Defensive	2,6,6	4,4,6	4,6,4	5,5,5

of power at sea, while country c has fine harbors but indefensible land borders, and so can influence the balance of power at sea but not on land. Each of the three countries must decide between a forward and a defensive military posture. The defensive position is cheaper and less likely to lead to war. The payoffs to the three countries are calibrated so that each receives a payoff of 5 when all choose cheap defensive postures, and each receives a payoff of 1 when all choose costly and risky forward strategies. However, when some choose forward strategies and others defensive, the country with the forward strategy can benefit. This can favor countries b and c, however, only if both choose forward strategies simultaneously: if only one of the countries does so, country a can concentrate its forces against that country and deprive them of any benefit from their enterprise.

This game has two Nash equilibria, one at the upper right where country a has a forward posture and b and c are defensive, and one at the lower left where country a maintains a defensive posture against forward postures by countries b and c. To say that the upper right cell in the table – strategy combination "forward," "defensive," "defensive" – is a Nash equilibrium is to say that no country can benefit by unilaterally deviating from it. Thus a shift by a would reduce its payoffs from 8 to 5; a shift by b would reduce its payoffs from 3 to 1, and similarly a shift by c. But if b and c were to make a coordinated shift from defensive to forward postures, country a's best response would be the defensive posture at the lower left, the other Nash equilibrium. Thus, the lower left equilibrium is *strong* in the terminology of Aumann (CGT4 pp. 287–324), and it is also *coalition-proof* in the terminology of Bernheim et al. (1987), while the equilibrium at the upper right is neither.

Strong equilibria and coalition-proof equilibria differ in detail, but both reflect the stability of the equilibrium against the formation of coalitions. If an equilibrium could be disrupted by a coordinated shift of strategies by some group of players, the members of the group are better off as a result *and the shift is to strategies that correspond to another Nash equilibrium,* then the first Nash equilibrium is rejected as unstable. Notice that the alliance between b and c, in the example, needs no enforcement – Nash equilibria are self-enforcing, and either country can only lose by deviating from it. That is why the phrase in italics, *"the shift is to strategies that correspond to another Nash equilibrium,"* is crucial: otherwise the new situation could not be sustainable without some enforcement. A *strong* Nash equilibrium is one that cannot be disrupted in this way, either because it is unique or because no group can benefit by a coordinated shift to another equilibrium. A *coalition-proof* equilibrium is one that either is strong or, if not, nevertheless is unlikely to be disrupted in this way, because any group shift to a new Nash equilibrium would lead to an equilibrium that would itself be unstable, in that a subgroup of the original group could benefit, at the expense of the rest, by a further coordinated shift.

This idea seems to have particular relevance for relations among sovereign states, as in the example. Sovereignty means that there is no enforcement of agreements, so relations among sovereign states are essentially noncooperative. Nevertheless, treaties and alliances can be effective and rather stable among sovereign states. This example suggests a reason why they can: the treaty or alliance corresponds to a Nash equilibrium, but there are other Nash equilibria that might otherwise be realized, that would be less advantageous to the allies or treaty partners. At the same time, and more crucially, the example indicates the limits of this possibility: if the terms of the treaty or alliance do not correspond to a Nash equilibrium, then in all probability they will not be kept.

As conventional solutions to a coordination game, alliances may be quite persistent. And in the real world, of course, there is more to it than this symmetrical model. A new alliance could well create irreversible changes in political and other circumstances that could make it impossible to go back to the old equilibrium. But we will have to defer any further speculation along these lines until Chapter 6, on the game in extensive form. (For an intuitive discussion of a real historical example, see McCain, 2014b, pp. 284–8, 310–11.)

4.6 REFINEMENTS

In addition, Game 4.7 illustrates a common problem of noncooperative game theory. When there are many solutions, how are we to choose among them? In Game 4.7, we excluded the equilibrium at the upper right on the grounds that a coalition of two countries could improve on it – it is neither strong nor coalition-proof. This would be an instance of a *refinement* of the Nash equilibrium. In general, a refinement is any assumption additional to the definition of the Nash equilibrium that allows us to reduce the number of Nash equilibria considered as solutions. We have already mentioned, in passing, another very important refinement: an evolutionarily stable strategy is a Nash equilibrium that is stable with an evolutionary dynamics.

In some games, we may find that one Nash equilibrium is better for *all* players than another equilibrium is: one equilibrium Pareto-dominates another. Certainly the dominated equilibrium would not be strong, but this dominance condition would give us an even more persuasive argument to rule it out. The fact remains that every pure-strategy Nash equilibrium has the consistent beliefs property – so that if each agent really believes that the other agent will choose a strategy that leads to the dominated equilibrium, he will do best to choose the corresponding strategy. Moreover, there are many proposed refinements of Nash equilibrium, and in some cases they conflict. Despite the large literature on refinements of Nash equilibrium, there is no refinement that assures us of a unique solution in every case, nor that is applicable in every case. As we will see in later chapters, some refinements may nevertheless be very important in particular cases.

4.7   EVOLUTIONARY GAMES

We have made passing references already to evolutionary interpretations of Nash equilibrium. Evolutionary models are an important branch of noncooperative game theory with broader implications.

As we recall, the evolutionarily stable strategy set is a refinement of Nash equilibrium, and the refinement might well be applicable to interactive decisions of human beings as well as to the evolution of species. As Friedman notes (1998), what distinguishes an evolutionary dynamics from other ways of looking at game theory is primarily a lack of foresight, in that agents do not anticipate or attempt to influence the future evolution of the decisions of others. Beyond that, the evolutionary dynamics might be linked to an otherwise completely rational decision process. However, one of the advantages of an evolutionary perspective is that we might instead approach game theory from the point of view of *bounded rationality*.

In neoclassical economics and classical game theory, individuals are supposed to be rational in the sense that they maximize their utility, profits or payoffs. This view has never been without critics, and Nobel Laureate Herbert Simon (Royal Swedish Academy of Sciences) was one of the most important and widely recognized critics. The phrase "bounded rationality" stems from him, and reflects his judgment that the maximization of utility, profits or other payoffs requires decisions that are far too complex to be within the cognitive capacity of real human beings. Instead, human rationality is *bounded* by the computational capacity of the human mind and brain. The typical *rational* activities of real human beings are the setting of targets and the search for alternative activities that are satisfactory according to those targets. This has been expressed by saying that real decision makers do not maximize but satisfice. To many economists, this is Simon's critique in a nutshell; but Simon was more than a critic and is known outside economics as one of the founders of artificial intelligence theory. In the tradition of artificial intelligence founded by Simon (with his collaborator Alan Newell: Newell and Simon, 1972; note also Simon, 1995), rational decisions are made by *heuristic rules*. These rules do not necessarily lead to maximization of profits or payoffs or anything else, but do lead to "satisfactory" results. Artificial intelligence envisions that real decisions are based on a large set of rules, a "knowledge base" or "rule base," not (as a rule) a single rule per decision – although a simple rule like "do what the boss says (if he is watching)" can play a great role in decision-making and the organization of human action. In general, though, even a naïve decision maker will have to decide what rule to apply – before deciding what to do – and the final decision can draw on a large number of rules working together. Experts in a field will have still larger and more complex rule-bases for their expert decisions. Think, for example, of a medical doctor's decision whether to recommend surgery to a patient: rules for the diagnosis of the condition, other rules that reflect the experience of the medical community about the usual results of alternative treatments for patients in particular categories, rules for the judgment of patient psychology and so on will all be brought to bear on the decision.

What perhaps needs stress is that bounded rationality is *not irrationality*. It is *rationality*, conceived in a way that reflects a realistic view of the limits of the human brain as a computer. (This is the way it is viewed by Simon and his followers if not by neoclassical economists.) Nevertheless it raises questions about the assumption of Nash equilibrium theory that agents infallibly choose the best response to the strategies chosen by others. Instead, we may link the theory of bounded rationality with that of evolutionary dynamics in game theory, yielding a more cognitively realistic noncooperative game theory (for example, Aumann, 1997; Gintis, 2007).

The evolutionary bounded rationality approach to game theory would begin by assuming bounded rationality. At a particular moment, then, agents choose their strategies according to a given set of heuristic rules. Some of these rules may generate strategies that better approximate best responses to the strategies chosen by others, and some generate strategies that approximate best responses less well. However, people learn, both from their own experimentation and the imitation of others, so that over time the rules that lead to poor strategies will be replaced by other rules that do better. In this way, decisions would tend, over time, to approximate the best-response decisions. Not all Nash equilibria will be stable with this sort of process. Those that are stable will be ESS and we can predict, tentatively, that actual decisions will be approximated by the ESS decisions.

In this view, the heuristic rules play the role in social evolution that genes play in organic evolution. (They are the "replicators:" Hodgson, 2002.) The pure strategies in the game play the role that individual plants and animals play in organic evolution: they are the interactors, and their interaction determines (via the payoffs that result) the tendency for the corresponding replicators (heuristic rules) to be replicated as a larger or smaller proportion of the population of rules as a whole. Thus it is the rules and decisions that evolve in this evolutionary view.

But is it *really* evolutionary? Those who have a turning to philosophy may question whether this sort of scheme really ought to be called evolutionary or whether evolution, which by definition advances no purpose or intention, can be applied to the decisions of human beings, decisions that are directed (within the limits of bounded rationality) to better realize the purposes and intentions of the human individuals. These issues are better avoided, so let us say that the scheme that combines ESS with bounded rationality is an *adaptive* game theory, based on the assumption that real human rationality is bounded but adaptive (compare Selten and Gigerenzer, 2001).

4.8   CONCLUSION

We find that a number of problematic cases in public policy can be traced to inefficient Nash equilibria with the games considered in strategic normal form. In this sense, we may say that noncooperative game theory is a powerful diagnostic tool for public policy. For now we reserve judgment as to whether it may also be helpful in prescription. Nash equilibrium models have proved a powerful tool of problem identification for public policy. While the Prisoner's Dilemma has commanded the central position in this perspective, a wide variety of other Nash equilibrium models may result in inefficient equilibria in the absence of some public action. In conflict situations, and some others, it may be rational for agents to be unpredictable, and to randomize their strategies; but in other circumstances, coordination and anticoordination games, randomization may be something to be avoided, and avoiding it may require some information not within the game itself. The Nash equilibrium in pure strategies has a property of consistent beliefs that helps to explain how a Nash equilibrium, once established, can be persistent even when other Nash equilibria may be more efficient. Extensions of these models, with large numbers of players, trial-and-error adaptive learning in place of ideal rationality, and lack of symmetry among players, may seem more "realistic" than the simpler Nash equilibrium models. Where there are two or more Nash equilibria, refinements may eliminate some as less plausible. Pareto-dominance, evolutionary stability, and resistance to disruption by coalitional shifts of strategy provide criteria for refinement in particular cases.

NOTES

1. For a case study see McCain et al. (2011).
2. I am indebted to my colleague Richard Hamilton for discussions that contributed to this example.

5. Correlated equilibrium

While Nash equilibrium is the central concept in noncooperative game theory, and has many applications, it is not quite the whole story. There are rival solution concepts and applications that are prescriptive rather than diagnostic. This chapter will discuss a major alternative: correlated strategy equilibrium.

A few years ago in New Zealand (Bray, 2003), telecommunications companies Teamtalk Ltd and MCS Digital Ltd were embroiled in a lawsuit. If both were to pursue their claims in a court of law, the legal fees would be great enough that both would be worse off. If one knew for certain that the other would pursue his claim, then the best response would be to abandon the claim, to avoid the legal costs. However, neither was certain that the other would withdraw, nor was willing to be the one to withdraw unilaterally. They agreed to settle the difference by arm-wrestling, the winner to take the asset and the loser to abandon his claim. On the face of it, this may seem an irrational procedure, but on more careful consideration it is quite rational. The two businessmen had arrived at a correlated equilibrium solution to their problem.

5.1 INTRODUCTORY EXAMPLE AND DEFINITION

Eastonia and Westoria are neighboring townships that share a business district on their border. Each is considering building a parking garage to serve the business district, but only one garage is economically feasible, so that if both build, both will be worse off. If one town builds, then the other has the further option to (1) improve their infrastructure to make it easier for people to make use of the parking garage in the nearby town, or (2) do nothing, retaining the reduced fees for on-street parking. Option (1) will have some cost but will capture a small part of the benefits of the parking structure in the form of increased business traffic. Most of the benefit of the parking garage will flow to the town that builds it, though. This is shown as Game 5.1, with 10 indicating the maximum benefit and other payoffs reflecting the assumptions above (Table 5.1).

This game has two pure strategy equilibria, each where one town builds and the other improves its infrastructure. There is also a mixed strategy

Table 5.1   Game 5.1: parking garage

Payoff order: Eastonia, Westoria		Westoria		
		Build	Improve	Do nothing
Eastonia	Build	−5, −5	10,2	3,1
	Improve	2,10	2,2	0,1
	Do nothing	1,3	1,0	1,1

equilibrium at which each town builds with probability 8/15 and improves with probability 7/15. A probability of zero is assigned to "do nothing," since any increase in the probability of "do nothing" will reduce the expected value payoff of the decision maker. Both of the pure strategy equilibria are efficient, but the benefits are very unequally distributed. The mixed strategy equilibrium is highly inefficient, however, with an expected value payoff of 2. In the absence of any custom or other signal to support a Schelling focal equilibrium in this case, as we have seen, the mixed strategy equilibrium would seem quite plausible as it applies to a case of complete ignorance.

In a *cooperative* arrangement, one could build and make a side payment to the other township so that both would share more equally in the benefits. Here, though, we are concerned with *noncooperative* arrangements. As we have seen, noncooperative equilibria must be self-enforcing, but can be randomized. The problem with the mixed strategy solution in this case is that it assigns a positive probability (0.28) to the lose-lose outcome at the upper left, where both towns build and lose 5. It also assigns a positive probability (0.22) to the outcome in the middle, where neither town builds. Suppose instead that the decision could be randomized in a way that would assign probabilities of zero to the upper-left and middle strategy combinations. This would assure that exactly one of the towns builds.

In fact, this is pretty easy to do. The two township supervisors could get together and flip a coin, with Eastonia building if the coin comes up heads, and Westoria building if the coin comes up tails. The result would be that the expected value for each township would be 6 – very much better than the 2 that would come from the mixed strategy equilibrium, and with the benefits equally shared in expected value terms. Then the township supervisor of Eastonia is choosing according to the rule "if heads then build else improve," and the Westorian supervisor is choosing according to the rule "if heads then improve else build." These rules are self-enforcing, in that the fall of the coin provides a signal for a focal equilibrium in Game 5.1.

This solution differs from a mixed strategy equilibrium in that the decisions are correlated. If Eastonia chooses "build" then Westoria chooses "improve" with probability 1, and if Eastonia chooses "improve" then Westoria chooses "build" with probability 1. Thus, it is called a "correlated strategy equilibrium," or more briefly, "correlated equilibrium." This concept, and much of the discussion here, is due to Luce and Raiffa (1957), although papers of Aumann (1974, 1987) have stimulated much of the interest in it. Aumann showed (1987) that correlated equilibrium can be a result of Bayesian learning, rather than conscious randomization and maximization, and several papers in the late 1990s (Foster and Vohra, 1997; Fudenberg and Levine, 1999; Hart and Mas-Colell, 2000) introduce adaptive (boundedly rational) procedures that lead to correlated equilibrium. Thus, correlated equilibrium is a very plausible adaptation in a game like Game 5.1.

The probabilities of the two Nash equilibrium pure strategy combinations would not necessarily be 50–50. Indeed, any probability between 0 and 1 would share the same self-enforcing property in Game 5.1. Figure 5.1 shows the range of correlated strategies for Game 5.1 as the probabilities assigned to the two Nash equilibria of the underlying game vary. In general there will be very many correlated strategies, and so it seems that we have only reproduced the problem of multiple Nash equilibria (though gaining some efficiency in the process). However, the equal probabilities do supply a cognitively salient focal point that could lead both agents to expect that correlated equilibrium rather than another. Moreover, in this case the principle of insufficient reason reinforces the equiprobable correlated equilibrium.[1] As we will see, there is some casual anthropological evidence that equiprobable solutions are very common. On the other hand, if the

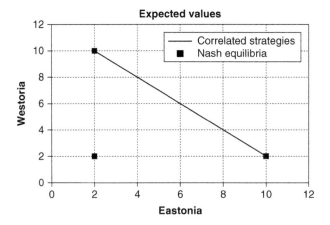

Figure 5.1   Nash equilibria and correlated strategies in Game 5.1

agents have an opportunity to agree on a signal, whether it is flipping a coin or arm-wrestling, they will also have an opportunity to settle the probabilities by negotiation.

There have been few applications to public policy. However, rationing by lottery has been fairly common in human history, and the military draft and similar arrangements may be seen as correlated equilibrium solutions. However, equal treatment in terms of expected values may not be seen as ethically or politically adequate. It may be that the objectives of public policy or the ethical standards from which public policy objectives are derived make reference to actual outcomes, so that these objectives and standards cannot be satisfied by unequal outcomes, even when all individuals have the same probabilities of being advantaged or disadvantaged. More generally, if we feel that different outcomes for different people require some justification, whether the justification is on the basis of different contributions, different needs, or different entitlements, differences resulting from a random mechanism may be seen as unacceptable. This consideration undoubtedly limits the application of correlated equilibria in public policy.

Thus, it may not be practical for the township supervisors in Game 5.1 to literally make their decision by flipping a coin. Yet they may nevertheless arrive at a correlated strategy solution to their interactive decision problem. Suppose they hire a consulting firm to study and compare the costs and benefits of the two alternative plans, for Eastonia to build and Westoria to improve and vice versa. Consulting the payoff table, we know that the net benefits are the same in either case, so any difference found by the cost–benefit study will be the result of errors in the study. We may suppose that the errors will be random and unbiased. Thus, when the township supervisors make their decisions on the basis of the cost–benefit study, they are carrying out a coordinated strategy solution! Of course, the cost–benefit study is likely to be a little more costly than flipping a coin. Still, it could be worthwhile as a politically respectable means of avoiding the impasse of the mixed strategy solution. We should note that the game is probably unrealistic in assuming that the two towns are perfectly symmetrical. Instead, we might want to assume that there actually are differences in the costs and benefits of the two proposals, but the town supervisors, not being specialists in cost–benefit analysis, do not know what they are. The payoffs are best-guesses, and the consulting firm is able to improve on them with more information and find real differences. However, the township supervisors are unable to anticipate that, and from their point of view, the subjective probabilities are 50–50. As Aumann has shown us, that is sufficient for the correlated equilibrium solution. When the New Zealand businessmen arm-wrestled to settle their impasse, each one probably thought that he was most likely to win, while an outside observer's subjective probabilities would perhaps be 50–50.

What the coin flip, the cost–benefit study and the arm-wrestling do is to supply a common signal to the two decision makers. The signal gives the decision makers the information they need to choose one of the plural Nash equilibria in pure strategies, and the decisions are correlated so that they will in fact correspond to one of the Nash equilibria in pure strategies. In the games we have considered in this section, it is clearly in the common interest of both decision makers to contrive such a common signal.

5.2   COORDINATION AND ANTICOORDINATION GAMES

Game 5.1, with its two Nash equilibria, has something in common with Games 4.5 and 4.6, the coordination and anticoordination games. Indeed, when the equilibrium of a coordination game is determined by custom, the custom might be considered as a signal that supports a correlated equilibrium. In some superspace of different possible histories, perhaps, driving on the left and driving on the right are equally probable – as witness the opposite customs in different countries. But, as we recall, anticoordination games present more difficult problems.

5.2.1   Stoplights as a Paradigm

The example of an anticoordination game is the intersection game – two cars approaching an intersection. Which will go through, and which will pause? Two Nash equilibria exist, where each car takes one of these roles. If they had time, the two drivers could get out and flip a coin to decide – but that would defeat the purpose of the exercise, which is to get through the intersection quickly. If the intersection is controlled by a stoplight, though, the car with the green light will go ahead and the car with the red light will stop. *This is a correlated strategy equilibrium.* The probability of getting a green light may be equal for the two, depending on when they arrive, or it may be unequal, if one of the streets is a major artery and so has longer green lights to accommodate heavier traffic; but, as we have seen, equal probability is not a requirement.

The traffic light is a fascinating twentieth-century innovation in the practice of interdependent decisions![2] It also illustrates the difference that makes anticoordination games more difficult than coordination games. For a coordination game, such as Game 4.5, the two decision makers receive the same signal, such as a customary practice, and that is sufficient so that they can coordinate their decisions. However, in the anticoordination game, different, correlated signals are required.

In a two-person game, given time, there is little difficulty in contriving this. However, the intersection game is a bit artificial, in that highway traffic is not really a two-person game. Rather it is a many-person game with drivers randomly matched to interact at intersections. In the individual matches, there will not be time enough for the drivers to get out and arm-wrestle. What is needed is a correlated set of signals for the entire population. The stoplights provide that correlated set of signals. But notice that the provision of common signals for this large population is a public good. It is no accident that stoplights are provided by government, although, in the early stages of motoring, some traffic direction was provided by private initiative.[3]

This defines a function of the public authority that has not been explicitly recognized, although it is implicit in some existing public activities such as the provision of stoplights and signage. As we will see it is also implicit in some aspects of economic policy. Perhaps explicit consideration of this role of the public authority will lead to innovations that can improve the results of private decisions in new ways.

5.2.2 Other Historical Instances

Lindow man is a famous mummy, the preserved cadaver of a Briton, seemingly a Celtic priest of pre-Roman times who was selected from among the members of a priestly community to be sacrificed to the gods and goddesses he served (Brothwell, 1987). His stomach contained the remnants of a burnt bannock. Interpreted in the light of Celtic tradition, it seems likely that he was chosen from among the community to be the sacrificial victim by the following procedure: those qualified to be the victim were required to draw a portion of bannock, a baked grain food, from a bowl without being able to see which fragment of the bannock they would take. One fragment was burnt, and the priest who drew the burnt bannock was sacrificed.

We may take it for granted that the priests felt that the sacrifice must occur; that the failure to make an appropriate sacrifice would be worse than death. At the same time we may take it equally for granted that each preferred not to be the sacrifice.[4] Thus, each faced an interdependent decision among three outcomes, in order of preference, another sacrificed, myself sacrificed, none sacrificed. If there are N priests this game has N Nash equilibria, each corresponding to the sacrifice of one of the priests. But how to decide among these alternatives? The priests adopted a correlated equilibrium, in which the probabilities attached to the different equilibria were roughly equal.

It may be that further discoveries will provide a different interpretation of this death. However, it will serve to illustrate the antiquity of correlated equilibrium solutions in practice, and the tendency to choose equiprobable schemes. Drawing straws, flipping a coin, and playing scissors–paper–stone[5] are all, as drawing the burnt bannock, arm-wrestling, and having an error-prone cost–benefit study done, means of assigning subjectively equal probabilities to two or more pure strategy Nash equilibria. A Methodist minister preaches that there are many instances of casting lots to make decisions in the Hebrew and Christian scriptures, dating from the Pentateuch (Lamar-Sterling, 2006). It was the "custom of the sea" that when sailors were cast away and starving, the victim to be cannibalized was chosen by casting lots or drawing straws (Hanson, 2001). What these examples show is that correlated equilibria with equiprobable assignments are widely known to nonspecialists in game theory and have been so, probably, for thousands of years. A rigorous anthropological study that would document the cross-cultural width and temporal depth of this knowledge would be of interest but is beyond the scope of this book. In short, correlated equilibrium is not mysterious, but just the opposite. However, we will see that an example due to Aumann has difficulties that we have not yet discussed, and that example has motivated most of the discussion of correlated equilibria in game theory.

5.3   DIFFICULT CASES

Now let us consider another example, with Eastonia and Westoria at loggerheads once again. This time there is a proposal for a cellular-telephone tower to serve both towns. There is a location on the border between the two towns that will serve them equally, but with minor inconvenience to both. In order to build at that site, both townships will have to agree to it. If just one approves the building of the tower, it will be built at a site within the approving township, and that township will be greatly inconvenienced. The other township will then get service of a quality almost as good as that from the borderline site, without any inconvenience at all. If neither approves, no tower is built, and cellphone reception will remain poor in both towns. As usual we will assign payoff numbers that agree with those assumptions so far as their relative magnitude is concerned, and not worry much about real units of measurement. Then we have Game 5.2 (Table 5.2).

This is a game discussed by Aumann in his 1974 paper that has been the source of most research on correlated equilibria. (Aumann refers only to "player 1" and "player 2" and offers no application in which the problem

Table 5.2   Game 5.2: a cellular telephone tower (Aumann's game)

Payoff order: Eastonia, Westoria		Westoria	
		Approve	Reject
Eastonia	Approve	6,6	2,7
	Reject	7,2	0,0

might arise, but the numbers have been chosen to agree with Aumann's; in any case they have the right relative magnitudes for our cell tower story.) Aumann's game has pure strategy Nash equilibria at the lower left and the upper right, and a mixed strategy equilibrium with probabilities 2/3, 1/3. The expected payoff of the mixed strategy equilibrium is 4 2/3, 4 2/3. Aumann writes (p. 72) "Consider now an objective chance mechanism that chooses one of three points A, B, C [that is, upper left, lower left, and upper right] with probability 1/3 each. After the point has been chosen player 1 is told whether or not A was chosen, and player 2 is told whether or not C was chosen; nothing more is told to the players." Then (p. 73) "the random device . . . is not at all difficult to construct. Given a roulette wheel, it is easy to construct electrical connections that will do the job." Aumann's example can be restated in terms of a more recent electronic technology, along the following lines.

We might construct a computer interface so that four possible events occur with probabilities determined by the program. The events are colored shapes shown on the screen: red triangle, red square, green triangle, and green square. The probabilities are 1/3, 1/3, 1/3, 0. Call the first three events, with positive probabilities, E_1, E_2, E_3. These probabilities are known to the players. However, the two players see separate screens, and Eastonia observes only the *shape* – triangle or square – while Westoria observes only the *color* – red or green. Thus, neither player actually knows what event has occurred. However if (for example) Eastonia observes a triangle, he can rule out one event, namely red square; that is, he can infer "not E_2." Similarly Westoria can observe red and infer "not E_3."

Suppose Westoria is known to choose according to the rule "if red then approve, otherwise reject;" that is, "if not E_3 then approve, otherwise reject." Let us call this contingent rule "rule A." Eastonia is considering adopting the rule "if triangle then approve, otherwise reject;" that is, "if not E_2 then approve, otherwise reject." Let us call this "rule B."

Case 1: suppose also that Eastonia observes square. He can infer that Westoria has observed red with probability 1 and so Westoria will choose approve, in accordance with rule A. In that case Eastonia's best choice is "reject" for 7 rather than 6. Case 2: suppose Eastonia observes triangle. He can infer that red square has not occurred (not E_2), but also that Westoria may play either accept or reject with equal probabilities, so that the conditional probability is ½. Again, this follows rule A. In this case the expected value of approve (that is, of following rule B) is $1/2(6 + 2) = 4$, while the expected value of reject is $1/2(7 + 0) = 3.5$. So Eastonia follows rule B and chooses approve. It follows that rule B is a best response to rule A, and similar reasoning establishes the converse.

If the decision makers coordinate their decisions in this way, probabilities of one-third each are assigned to the upper left, lower left, and upper right cells, and zero to the lower right cell. The overall expected value for each player is $1/3(7 + 6 + 2) = 5$, so that this contingent strategy dominates the mixed strategy Nash equilibrium. Suppose instead that probabilities ½ had been assigned to the two pure strategy equilibria as in a stoplight equilibrium, and zero to "approve, approve." Then the expected value for each town would be 4 ½ – so the correlated equilibrium can do better than simply a weighted sum of the pure strategy Nash equilibria. This is Aumann's key conclusion.

Rules A and B provide a set of strategies (contingent on the private signals from the two computer screens) that are best responses to one another, thus need no enforcement; are not Nash equilibria because the strategy choices of the agents are not independent but partially correlated; and are more efficient than any Nash equilibrium or probability mixture of Nash equilibria. It is a correlated strategy equilibrium, like the ones we have seen in coordination and anticoordination games; but the game is not a coordination or anticoordination game and is more difficult in two senses.

First, in Aumann's game, the correlated strategy equilibrium is not efficient. A slight increase in the probability of "approve, approve" at the expense of the other two (Nash equilibrial) outcomes will increase the expected value payoffs of both players. However, there is a limit to how far this can be taken. If either township is certain that the other will approve the project, then its own best response is to reject. In Game 5.2, if the probability of the red triangle ("approve, approve") is more than one half, then rules A and B are no longer self-enforcing. And the correlated equilibrium mechanism can assure the two individuals an expected value no greater than 5.25 each, whereas an enforceable cooperative agreement would assure them of 6 each. For this reason, also, signals must not be public, but private, and cannot be perfectly correlated, but must be imperfectly correlated. Put otherwise, the effectiveness of the correlated equilibrium mechanism in this case requires that each agent be at least somewhat ignorant of the other's plans – but ignorant only to just the right degree! The provision of private, imperfectly but appropriately correlated signals to different agents in a game is a difficulty that does not arise for coordination or anticoordination games. A further difficulty is that the set of correlated equilibria is an open set, so that the optimum correlated equilibrium is undefined.

5.4 SUNSPOT EQUILIBRIA AND ECONOMIC POLICY

Stanley Jevons (1835–82) was one of the great founders of neoclassical economics, and also pioneered statistical work in economics and made contributions to symbolic logic, including a prototype computer[6] (New School History of Economic Thought Website, 2007). Before he became an academic economist and philosopher, he worked as a geologist and was an amateur meteorologist. He proposed that sunspots might be a cause of business cycles. By the late twentieth century "a sunspot" came to refer to any fluctuating quantity that might seem to have a correlation with business activity but could have no causal influence. A literature of the late twentieth century suggested that such "sunspot" variables might nevertheless partly determine economic activity.

By this time it was generally known that "rational expectations" equilibria in macroeconomics would not in general be unique, but that, indeed, a macroeconomic model could have many equilibria. In overlapping-generation models Azariadis (1981) and Cass and Shell (1983) showed that the agents might correlate their decisions with a "sunspot" variable, so that a "sunspot equilibrium" might be observed – even though the sunspot would have no causal effect on any real economic variable. A quite large literature followed, of which the following are just a few suggestive contributions. Prima facie, sunspot equilibria seem to resemble correlated equilibria in noncooperative game theory. However, Maskin and Tirole (1987), with a fixed-horizon model, obtain largely negative results, in that the conditions for a correlated equilibrium to correspond to a sunspot equilibrium are quite limited. On the other hand, Peck and Shell (1991) provide an example with imperfect competition in which the sunspot equilibria are correlated equilibria of a market game. Chatterjee et al. (1993), in a two-sector over-lapping-generations model, argue that complementarity can result in multiple equilibria and fluctuating sunspot equilibria. The following example is suggested by, but not necessarily an instance of, these contributions.

We will begin with a two-person game of market entry. Firms A and B can each choose between immediate entry (go) and postponement of entry to the next period (postpone). While their operations are complementary, in that each supplies a cheap or highly effective input to the other, their

Table 5.3   Game 5.3: complementary market entry

Payoffs: A, B		Firm B	
		Go	Postpone
Firm A	Go	10,5	0,0
	Postpone	8,8	5,10

technologies are different. Firm B uses a roundabout method that will be more effective in the second period, after a preparatory phase in the first period, so postponement by both firms favors Firm B, and this will be expressed by payoffs 5,10, where the first payoff is to Firm A. Firm A relies on a wasting resource which is more available in the first period, so the case in which neither firm postpones is favorable to Firm A, expressed by payoffs 10,5. If B goes ahead with entry and A postpones, B's first period production increases the stock of input to Firm A that offsets the wasting of its resource that would otherwise occur, so both benefit to some extent from complementarity in both periods and this is expressed by payoffs 8,8. However, if B postpones and A does not, neither benefits from complementarity in any period, expressed by payoffs 0,0. The resulting Game 5.3 is shown as Table 5.3.

This game has three Nash equilibria: pure strategy equilibria in the upper left and lower right, and a mixed strategy in which A assigns a probability of 0.714 to postponement and B assigns a probability of about 0.286 to postponement. In this mixed strategy equilibrium, each firm has an expected value payoff of 7.14.

Now, suppose that the presidents of both firms observe sunspots, and that the probability that sunspot activity is greater than its mean is just 0.5. Suppose each businessman believes that when sunspot activity is above mean, times are good for new entering firms, but when sunspots are below average it is best to postpone entry. Then both enter with probability ½ and both postpone with the same probability, ½. In that case, they have a correlated equilibrium with expected value payoffs of 7.5. Moreover, a market for financial securities, that is "contingent claims," can play a role in this correlated equilibrium. Suppose Firms A and B sign a contract that specifies that A will pay 2.5 to B if sunspot activity is greater than average, and B will pay 2.5 if not. Then, by playing the correlated equilibrium, the two agents each obtain a payoff of 7.5 with certainty.

We might make this a basis for a larger game. Suppose that at each time $t = 1, 2, \ldots, q$ new firms of each type come into existence. The firms are matched in complementary pairs, perhaps because of different locations, with payoffs determined as in Game 5.3. Now suppose that above-average sunspots in periods t and $t + 1$ are followed by below-average sunspots in period $t + 2$. Firms that have entered in period t cease to exist in period $t + 2$, and the potential new firms in period $t + 2$ postpone their entry until period $t + 3$. Thus, $2q$ firms of each type are active in period $t + 1$, but period $t + 2$ is a recession period with only q firms of each type active. Now suppose that sunspots are again above average in period $t + 3$. Potential new firms from both periods $t + 2$ and $t + 3$ enter in period $t + 3$, and we have a recovery, with the number of active firms returning to $2q$ of each type. Thus we observe "business cycles" correlated with sunspots although sunspots have no causal influence on economic variables.

But it is possible to improve on the sunspot equilibrium in this instance. Suppose instead that a disinterested third party gives each firm an instruction either to enter or to postpone entry. The third party randomizes her instructions, instructing both to enter with probability ¼, both to postpone with probability ¼, and A to enter and B to postpone with probability ½. The firms are also instructed to keep their instructions confidential. Suppose Firm B is instructed to enter without postponement. Computing the conditional probability that Firm A also will enter without delay as $(\frac{1}{2})/(\frac{1}{2} + \frac{1}{4}) = \frac{2}{3}$, Firm A's expected value from following the instruction is 7, while the expected value of acting against the instruction, that is, postponing entry, is 6.7. Thus Firm B's best response is to follow the instruction. Moreover, Firm B has an incentive to keep his instruction confidential, as he is also instructed, since if he were to reveal that he is to enter immediately, A's best response would be to do the same, reducing B's payoff to 5 with certainty. Suppose B is instructed to postpone. In that case, he has nothing to lose either by following the instruction or by keeping it confidential, since he knows certainly that Firm A's instruction is also to postpone. Firm B can realize his maximum payoff by following this instruction without revealing it and revealing the instruction will not improve on that.

Suppose Firm A is instructed to postpone entry. Computing the conditional probability for Firm B's instructions as before, Firm A also finds that its expected value is greater from carrying out the instructions than from deviating from them; and moreover that she is better off to keep the instruction confidential, since if Firm B knew with certainty that firm A would postpone, Firm B would postpone, reducing Firm A's expected value from 7 to 5.

All in all, then, the third party's randomized instructions are self-enforcing, and since the overall result of following instructions is an expected value payout of 7.75, the instructions Pareto-dominate both the previous correlated equilibrium, with a public "sunspot" signal, and the mixed strategy Nash equilibrium.

In these examples, the signal that coordinates decisions for the two firms is an astronomical phenomenon or an instruction from an anonymous third party. As we noted in the discussion of stoplights, provision of a coordinating signal might be a function of the public authority, as the provision of a public good. To what extent, then, might the provision of coordinating signals for the macroeconomy be a public function? Indeed, arguably, it already is. The importance of the "announcement effects" of policy announcements by the Federal Reserve System in the United States has long been known (Waud, 1970). It has been suggested (Stein, 1989) that the Fed deliberately practices "cheap talk" as a means of influencing economic activity, though the generality of the reasoning has been challenged (Conlon, 1993). Federal Reserve announcements – and similar announcements from other public bodies – could play the role of public "sunspots" in determining a correlated equilibrium in the "game" of macroeconomics. In the example with private signals, the "third party" giving the instructions sounds a bit like an economic planning agency.[7] This is not to suggest that any economic planning agency has ever, in practice, had the knowledge necessary to construct a self-enforcing plan, using randomized strategies, as that example proposes. It is an open question whether this might be practicable in some future institutional context. The conclusion we may draw is that the public provision of coordinating signals is a public function about which very little is known, and that merits extensive future research.

5.5 PLURAL NASH EQUILIBRIA AND THE RATIONALITY POSTULATE

It has been argued (Hargreaves Heap and Varoufakis, 1995; Coleman, 2003) that the multiplicity of Nash equilibria impeaches the rationality postulate basic to game theory (as well as neoclassical economics). When there are plural equilibria, the agents cannot determine their decisions simply through rational procedures. At best, custom and convention come into their decisions, and at worst, the decisions are simply indeterminate. This is said to discredit the rationality postulate. This critique goes too far, however.[8]

In the case of coordination games, such as Game 4.5, custom can indeed be the decisive determinant of the decisions. But the best-response principle is an explanatory principle without which custom might not determine the decisions in these cases. In Game 4.5, for example, the key point is that following custom is a best response; if people do not predictably follow their best response in Game 4.5 then the custom itself loses its predictive role. The nonrational alternative would be to suppose that people mechanically follow custom regardless of whether it is a best response or not. This in turn would mean that in Game 4.1, the pollution game, a custom of nonpollution would be sufficient to assure the efficient outcome. Game theory predicts the opposite, setting limits to the cases in which custom may be decisive. Indeed, the weakness of custom in limiting the deployment of polluting technologies does seem to contrast with the power of custom in determining that Britons drive on the left side of the road and Americans on the right. In any case, this is an elaboration, not a failure, of the rationality hypothesis.

The case is more difficult in an anticoordination game, such as Game 4.6. In such a game a uniform signal, such as a convention along the lines of "drive on the left," will not do, so that the decision is all the more likely to be indeterminate. Yet, in fact, a more complex custom may resolve the decision. If the participants in the game are ranked, so that the person of higher rank is given precedence, then the information as to which agent is of the higher rank makes the decision determinate. A difficulty is that the hierarchy of rank must be complete: it must be that every match is between two agents of different rank. Rank in military organizations illustrates this. Combat generally involves coordinated action but differentiated missions and objectives, with great and widely different risks. In battlefield conditions, coordination is crucially important and indeterminate decisions can result in disaster. Thus, it is crucial that some specific person is able to make authoritative decisions, and far less important to make fine calculations about who is best suited to make them. The system of military ranks, with persons of equal rank subordinated by seniority or even age, is well suited as a customary solution to this problem.

On its face, it may seem that the correlated equilibria make the case worse, because they are far more numerous than Nash equilibria in many games. When there are two or more undominated Nash equilibria, there is a continuum of correlated equilibria. However, this is misleading. When the two businessmen agreed to settle their difference by arm-wrestling, the probabilities (whatever they may have been) were probabilities they agreed on, and their agreement specified a single one of the infinitely many possible allocations of probabilities among the two different pure-strategy Nash equilibria. The same is true when the township supervisors of Eastonia and Westoria agree to base their decision on a cost–benefit study, although each is certain that his own town is the best choice, and when two drivers at an intersection happen unpredictably to arrive when the light is red one way and green the other, and of Celtic priests deciding who is to be sacrificed.

It is true that there may not be time to come to agreement on the probabilities or on a mechanism, such as arm-wrestling or drawing the short straw, so that a correlated equilibrium may simply not be available as a means of resolving an indeterminate Nash equilibrium. The meeting of two cars at an intersection has been given as an example – and probably most readers of this book have experienced impasses of this kind. It is also true that, in a game like Game 5.3, Aumann's game, there is potential of increasing efficiency beyond what an average of the Nash equilibria can support; but this potential can only be realized with private, imperfectly correlated signals, which may be difficult or costly to arrange. But it is not the rationality postulate that underlies these failures – it is the lack of sufficient time or communication.

The widespread observation of equiprobable mechanisms suggest that, in the absence of clear reasons against them, equiprobable correlated equilibria will usually occur when there are multiple undominated Nash equilibria. That equal probabilities are cognitively salient and determine a Schelling focal point, and require little knowledge but are consistent with the principle of insufficient reason, further points in the favor of the prediction of an equiprobable solution. If there is a custom or convention that supplies a solution, then custom will take precedence over an equiprobable chance mechanism. In the rest of this book, cases of multiple Nash equilibria that cannot be resolved otherwise will be assumed to lead to equiprobable correlated equilibria.

5.6   CHAPTER CONCLUSION

Nash equilibria do not exhaust noncooperative game theory. Correlated equilibria, with strategies that are randomized but not independent of one another, expand the set of possible noncooperative solutions, but also resolve many of the puzzles about games with two or more undominated Nash equilibria. It should be added that the correlated equilibria can be thought of as Nash equilibria in enlarged games that include such things as arm-wrestling or drawing straws, with strategies contingent on events in those stages. It is the fact that they are Nash equilibria of enlarged games that makes them self-enforcing. Thus, the correlated equilibria are a structure built on the foundation of Nash equilibria. They cannot supplant Nash equilibria in noncooperative game theory. Public policy applications of correlated equilibria are largely unexplored. Nevertheless, these studies enrich our understanding of the theory of noncooperative games in strategic normal form in important ways and their role in public policy deserves extensive research.

NOTES

1. The "principle of insufficient reason" is at best controversial. If we reject it entirely, though, then multiple equilibria are not likely to be much of a problem – people will always have some grounds for thinking one equilibrium more likely than another, and the set of equilibria collapses to that one that is thought more likely. Schelling's (1960) discussion follows very much that line. On the other hand, if we take seriously the idea that the agents have no reason to expect one equilibrium rather than another, we might express that by saying that their information is minimal; and the minimum-information condition for an exhaustive set of exclusive alternatives is that the alternatives are equiprobable. (See McCain, 1972, for discussion and an application.)
2. The invention of the traffic signal has been widely attributed to Garret Morgan, an African–American inventor (Famousinventors.com, 2006) although there has been some controversy on the point. It does seem clear that Morgan was, at least, *an* inventor of a device to control traffic at intersections, bringing about a correlated equilibrium solution to the anticoordination game of intersection traffic.
3. Keystone Automobile Club (1927).
4. It would make little difference if each wished the honor of being sacrificed, provided that the sacrifice must be unique.
5. Since the only Nash equilibrium of scissors–paper–stone is a mixed strategy equilibrium with equal probabilities, this method uses a game with a unique solution to generate the probabilities to choose among the equilibria of a game with plural Nash equilibria.
6. These comments were derived from the New School History of Economic Thought Website in 2007. Unfortunately, the New School University no longer provides that service.
7. McCain (1991) proposed a model of economic planning for market economies based on a coordination game model which in turn was derived from Rosenstein–Rodan (1943). This model was extended to asymmetric information in McCain (1985). The 1991 paper, though preliminary, was long delayed in publication. McCain's model assumes only public signals, however.
8. I do not mean to suggest that there are no valid criticisms of the rationality postulate, as discussions at Chapter 3 and Chapter 9 may illustrate.

6. Noncooperative games in extensive form and public policy

In Chapters 4 and 5, our focus was on noncooperative games in strategic normal form. While (as von Neumann and Morgenstern showed) all games in extensive form can be represented in strategic normal form, to do so in general we may have to be careful to specify strategies as contingency plans. Thus, the strategic normal form will apply most naturally and with the best intuition to games in which simultaneous choices of behavior strategies must be made, such as the Prisoner's Dilemma. Conversely, when some decisions must in fact be made before other decisions are made, so that subsequent decisions are made with knowledge of the earlier decisions, the game represented in extensive form may be more natural and intuitive. In this chapter we focus on the game represented in extensive form.

6.1 SUBGAME PERFECTION AND TREMBLING HANDS

Recall Game 2.2, Figure 2.1 in Chapter 2. We should notice that the decision by Firm A, to accommodate or retaliate, is a subgame in this game. Accordingly, we can define a *behavior strategy* locally at this decision point. The behavior strategy is just to accommodate or to retaliate, without specifying any conditions as to what previous decisions might be made. (Such conditions would be trivial in this case anyway.) In the spirit of Nash equilibrium theory, we might suppose that Firm A will choose the behavior strategy that leaves it with the larger payoff. This is "accommodate" for a payoff of 2 rather than 1. Moreover, the potential entrant, Firm B, can anticipate this. Therefore, Firm B expects that the payoff from the behavior strategy "enter" pays 1 while the behavior strategy "don't" pays 0, and accordingly Firm B chooses "enter." Thus the noncooperative solution to this game would seem to be "enter, accommodate."

Four comments should be made on this reasoning.

First, it is an example of *subgame perfect Nash equilibrium*, a concept that is now central to the analysis of games in extensive form. A subgame perfect Nash equilibrium is a sequence of behavior strategies that (1) is a Nash equilibrium in behavior strategies in the game as a whole, and (2) is also a Nash equilibrium in every subgame. In this case, we have just one proper subgame, and that is Firm A's decision whether to retaliate or accommodate. The fact that Firm A chooses the behavior strategy that maximizes its payoffs at that point means that we do indeed have a Nash equilibrium in this subgame. That Firm B maximizes its own payoff based on anticipation of that decision means that each firm is choosing its best response to the other's strategy (sequence of behavior strategies); we have a Nash equilibrium in the game as a whole.

Second, the example illustrates an algorithm for finding subgame perfect Nash equilibria. The algorithm is called "backward induction." In this case, notice, the first step is the last decision to be made, resolved by determining a Nash equilibrial behavior strategy as if the subgame stood alone. We then treated the first decision as a "reduced game" in which the payoffs were 0 for "don't" and 2 – the equilibrial payoff in the first step – for "enter." The Nash equilibrial decision for the "reduced game," "enter," then completes the subgame perfect Nash equilibrium. For more complex games and in general, the algorithm would be as follows: (1) Among all subgames, determine those that are *basic.* A basic subgame is one that has no proper subgames within it; in Game 2.2, Firm A's decision is the only basic subgame. (2) Determine the behavior strategies that constitute a Nash equilibrium for the basic subgames. (3) If the Nash equilibrium is unique, form the *reduced game* by eliminating the basic proper subgames, replacing the basic proper subgames by their equilibrial payoffs. If the Nash equilibrium for a particular subgame is not unique, replace the subgame by one or another set of equilibrial payoffs. (4) Repeat until the reduced game is the first decision to be made, and determine the Nash equilibrial behavior strategies and payoffs for that decision. (5) The sequence of behavior strategies are then the subgame perfect Nash equilibrial behavior strategies, and the payoffs yielded by this sequence are the subgame perfect equilibrium payoffs.[1] If one or more of the equilibria determined at stage 3 are non-unique, then the subgame perfect Nash equilibrium is non-unique.

Third, another way to express the result in this analysis is to say that the threat of a price war in this case is *incredible.* This recalls, yet again, Nash's comment that a threat is often something that a person would not want to do for themselves, and that is the case with respect to the price war in Game 2.2. In general, in noncooperative game theory, a threat is *credible* only if it is subgame perfect. If the threat is part of a subgame perfect equilibrium sequence of behavior strategies, then it is Nash equilibrial in the subgames of which it is a part, and if so then it is an exception to Nash's comment – it is something the person would want to do for itself.

Fourth, the application in this case is itself very important for public policy and economics. It supports the argument that market entry is irrepressible in a market economy without government restrictions. Since entry tends to increase price competition and price competition in turn tends to induce efficient pricing and resource allocation, this would be an element of an argument for free market policies. On the other hand, if in a special case market entry were to have negative consequences, it could be an element of an argument for public policies that would restrict entry. Patent rights would be an instance of the special case.

Although we are concerned with the representation of the game in extensive form, it will be helpful here to digress on the strategic normal form. Recall the normal form of this game, Table 2.2 in Chapter 2. Notice that this game has four Nash equilibria: "don't" with strategies 1 and 4, and "enter" with strategies 2 and 3. The latter two correspond to the subgame perfect Nash equilibrium, since both require Firm A to choose the behavior strategy "accommodate." But the other two formally are Nash equilibria as well.

Is there any basis to exclude these equilibria? We do notice that strategies 1 and 4 are weakly dominated for Firm A. A strategy is *weakly dominated* if there is another strategy the payoff to which is never less, and is greater for at least one strategy that the other player might choose. Since behavior strategy "accommodate" pays 2 if Firm B enters, and 5 if Firm B does not, the contingent strategies leading to "accommodate" weakly dominate those leading to "retaliate." Indeed, we see that this game has only three distinct outcomes: price war, accommodated entry, and continued monopoly. Behavior strategy "don't enter" always leads to the same outcome, therefore to the same payoffs. In general, when we translate a game in extensive form into a game in strategic normal form, we will find that there are many fewer outcomes than strategy combinations, since many different combinations of contingent strategies will lead to the same basic subgames, and therefore to the same outcomes. Thus, weakly dominated strategies are likely to be quite common in games in extensive form. The question thus becomes: is there any basis to exclude equilibria that are based on weakly dominated contingent strategies?

In 1975, Selten (CGT, pp. 317–54) introduced a refinement of Nash equilibrium called the *trembling hand* equilibrium: suppose there is some small positive probability that a player will fail to choose his best-response strategy, so that the player will choose any other specific behavior strategy instead. This is a *perturbed* game. We can define equilibrium for the perturbed game in the usual way, with the expected values of payoffs determining the best responses. The equilibria of a perturbed game may differ from those of the original game. Now define a sequence of perturbed games in which the probability of errors approaches zero in the limit. Selten shows that the limit of the equilibria of such an (appropriately constructed) sequence of perturbed games is an equilibrium of the original game, but not all equilibria are the limits of such sequences. Those equilibria that are the limits of such sequences are *perfect equilibria*. This means that equilibria are excluded if they depend on behavior that is rational only on the assumption that other players are themselves perfectly rational.

In a perfect equilibrium, a Nash equilibrium is realized in every subgame of the original game, including of course the game itself. Thus a perfect equilibrium is, in particular, subgame perfect. In Game 2.2 revised, for example, suppose that the probability that Firm B chooses the "wrong" strategy is p. In such a case the payoff of contingent strategies 1 and 4 is $2p + 1(1 - p)$. The payoff to strategies 2 and 3 is $5p + 2(1 - p)$. Clearly the second is larger for any positive p, so strategies 1 and 4 are not best responses in any perturbed game. Consequently, the subgame perfect equilibrium of Game 2.2 is the only perfect equilibrium of Game 2.2 revised.

But the perfect equilibrium can also be applied to games that do not have subgames. As an example, Selten discusses the Horse game (Figure 2.3, Game 2.3, Chapter 2). As in Chapter 2 we will encode the behavior strategies as follows: for Firm A, "License" is R1, "Don't" is L1, for Firm B, "Don't" is R2, "Enter" is L2, and for Firm C, "License" is R3 and "Don't" is L3. Notice that for this game, L1R2R3 is an equilibrium: but it is so only because Firm B does not get an opportunity to play at all. If Firm B were to get an opportunity to play, he would know that Firm A had not played L1 but R1 and, on an expectation that Firm C would play R3, Firm B's best response is not R2 but L2. This is unreasonable, Selten argues, writing (p. 328) "Player 2's choices should not be guided by his payoff expectations in the whole game but by his conditional payoff expectations" at decision node B. In fact, L1R2R3 is not a perfect equilibrium. In Game 2.2 revised, the Nash equilibria with contingent strategies 1 and 4 can be excluded because they are not perfect equilibria. As we noted, they are formally Nash equilibria, but their exclusion is very much in the spirit of Nash's noncooperative game theory. Retaliation is a threat strategy, and would not be "something A would want to do, just of itself." The term "perfect Nash equilibrium" is quite apposite: rather than restricting, Selten has perfected Nash's reasoning.

6.2 PRAGMATICS: PROBLEM SPECIFICATION

As before, one of our concerns is with problem identification. Accordingly, we consider some cases in which the game in extensive form helps us to specify a problem of interactive decision-making that may be relevant for public policy.

6.2.1 Ulysses and the Sirens

We recall that the entrenched monopolist in Game 2.2 was unable to prevent the entry of new competition, because the threat of a price war was not subgame perfect. Yet the entrenched monopolist might not be entirely helpless. The monopolist might consider a large-scale investment that would increase both its production capacity and its costs, creating a situation in which the restricted output corresponding to accommodation of the new entering firm would be less profitable, and the increased production incident on a price war more profitable. If it anticipates competitive entry (for example, following deregulation) the firm might undertake such an investment, in the hope that the entry would be prevented. This is *strategic investment to deter entry,* and Game 6.1 is an example. (See Figure 6.1; the numbers in parentheses will be explained below.) As before, the first payoff is to Firm B.

In Game 6.1, if Firm A decides not to invest, then we have Game 2.2, but if Firm A decides to invest, we have a different subgame. To solve this more complex game, we again identify the basic subgames, and they are A's second round of decisions. Both are basic. For the lower one, we already know that the perfect behavior strategy is "accommodate." For the upper subgame, however, "retaliate" is the perfect response. Thus, Firm B can anticipate that the behavior strategy "enter" will pay 1 in the lower subgame but −1 in the upper. "Enter" is a perfect behavior strategy in the lower subgame but not the upper. Anticipating all this, Firm A expects that investing will lead to profits of 4 while not investing will lead to profits of 1. Thus, the subgame perfect sequence of behavior strategies is "invest," "don't enter."

The investment may or may not be efficient. To know the answer to that, we need to know more about the benefits to customers. In economics, the *consumers' surplus* measures the net benefit to the buyer from buying at a particular price. The consumers' surplus plus the total profits of the two firms measures the net social benefit for this industry, in the absence of externalities. For this example, the numbers in parentheses represent consumers' surpluses corresponding to the different degrees of price competition and output capacity in the various possible outcomes of the game. The largest total, 8 on our arbitrary scale of measurement, occurs if Firm A does not make the investment, Firm B does enter, and entry is accommodated. In that case, the consumers have the benefits of both expanded production capacity and increased price competition. By contrast, in the

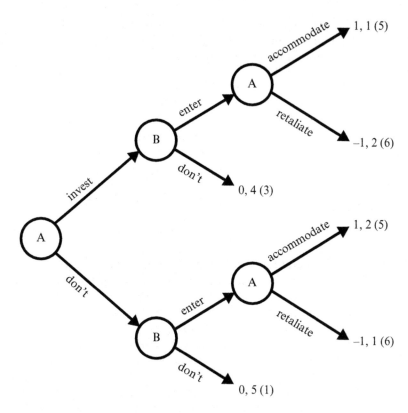

Figure 6.1   Game 6.1: strategic investment to deter entry

subgame perfect equilibrium, consumers benefit from increased capacity but not from increased price competition, so the total, 7, is less. Efficiency is improved – without either entry or new investment, at the bottom of the diagram, total benefit is 6 – but the resulting outcome is not fully efficient because of the entry-limiting investment.

The key point is that the investment has been successful in deterring entry. In Game 2.2, Firm A was unable to deter entry because the threat of a price war was incredible: Firm B could anticipate that Firm A would not undertake such an unprofitable step. If Firm A were able to do so, it would choose the best response behavior strategy of accommodation. However, by making the strategic investment, Firm A has deprived itself of the opportunity to accommodate profitably.

This illustrates a more general point that emerges from the study of the game in extensive form. In some cases, an agent may be better off with fewer opportunities, fewer options. Schelling (1960) stressed that in order to bind others, we might have to find a way to bind ourselves. This is what Firm A has done in Game 6.1. Elster (1977) drew the analogy to Ulysses having himself bound to the mast so that he could hear the song of the sirens – and not succumb to it.

By considering a strategic investment to deter entry, Firm A has *nested* Game 2.2 within a larger game. Since Game 2.2 is a subgame of Game 6.1, it is an *imbedded* game. That is important, since it means that the subgame perfect solution of Game 2.2 is perfect also in the context of the nesting Game 6.1. That is to say, (for purposes of noncooperative game theory) we can analyze the imbedded game as a game in its own right, expecting that its equilibrium will be equilibrial also within the larger game – while that might not be true for a game that is nested but not imbedded in the larger game.

6.2.2 Agency

A large body of literature in economics and game theory has grown up with respect to relations between a principal[2] and an agent. The principal sets the conditions for the decisions taken by the agent. There are some aspects of the agent's activity, such as effort, that the principal is unable to observe. It is this that makes the relationship (at least partly) noncooperative. Agency models are also more general than the phrase may suggest. The agent may sell a home on behalf of the owner, but may be a corporate executive when the principals are shareholders, or may be a professional and the principal a client, and so on. As usual we will illustrate the idea with a simplified (and nonmathematical) example; the agent will be a lawyer and the principal a client.

For the example, there is a third player: chance. The lawyer can pursue the lawsuit with great or slight effort, and these are her strategies. Chance can provide good or bad conditions for the lawsuit, at random with equal probabilities ½. If the lawyer makes a great effort and chance is favorable, the lawsuit yields 14 to the plaintiff. (As usual the reader may add as many zeros as seem appropriate.) For the lawyer, great effort is equivalent to a deduction of 2 from her fee, while slight effort is equivalent to a deduction of 1. If the lawyer makes a slight effort and chance is unfavorable, the lawsuit yields only 2. If the lawyer makes great effort and chance is unfavorable, *or* if the lawyer makes slight effort and chance is favorable, the lawsuit yields 6. Since the client cannot observe either effort or the random variate, he will not know whether the intermediate outcome is a consequence of slight effort or of adverse chance. As a result, the payment for legal services cannot be conditioned on effort. The client is considering whether to pay the lawyer a flat fee of 3 or a contingent fee of 1/3 of the award.

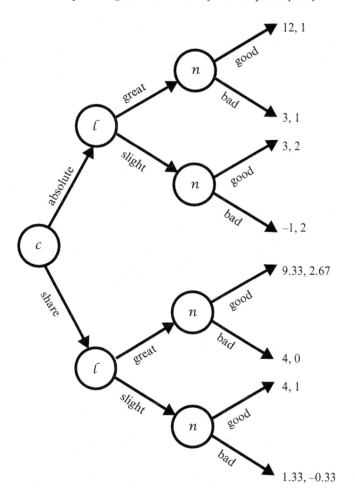

Figure 6.2 Game 6.2: agency

The resulting game in extensive form is shown as Game 6.2, with c denoting the client, l the lawyer and n (for nature) chance (Figure 6.2). The first payment is the net benefit to the client, the second to the lawyer. Chance "decisions" are not subgames (since chance has no intentions) and the basic subgames are the lawyer's decisions. With a fixed fee of 3 she is always better off making slight effort, but with the contingency fee she can anticipate an expected value of 1 1/3 with great effort, but 1/3 with slight effort, so in the lower part of the diagram, the lawyer will choose to make a great effort. Accordingly, the client will face an expected value of 1 (net of the fee of 3) in the case of a flat fee but 6 2/3 in the case of a contingency fee. The subgame perfect equilibrium is for the client to offer a contingency fee and the lawyer to make great effort.

In Williamson's informal work on the economics of the firm and similar cooperative arrangements (for example, 1964), he put some stress on "opportunism," arguing that many seemingly inefficient contractual arrangements could be explained as means of avoiding or limiting opportunism. He was sometimes criticized by economists who suggested that, after all, opportunism is nothing more than self-interest. In the agency example, suppose that the lawyer promises to make a great effort in exchange for a fixed fee. This would generate an expected value of 7.5 for the client, so that, if she trusts the lawyer, the client would accept that offer. If, however, the lawyer nevertheless makes slight effort, we would say, in a sense of everyday usage, that the lawyer had acted opportunistically, and that the choice of a contingency fee is made to avoid the consequences of opportunism, very much along the lines of Williamson's thinking. It is true enough that opportunism is *noncooperative* self-interest, but also that opportunism is not identified by the character of the behavior alone but also by the sequential structure of the game.

6.3   IMBEDDED GAMES

In Chapter 2, a game was said to be imbedded in another if the first game is a proper subgame of the second. The imbedded game can be studied as a stand-alone noncooperative game. Since we cannot represent the universe as a single game, this is essential to any applied game theory. In particular, it is important for public policy applications. If we think of the public sector as setting the "rules of the game," then private sector (interdependent) decisions are imbedded within the game of public policy determination. We will illustrate the point, as usual, by example.

Let us return to an example from Chapter 2: two agents own property on a river. If Joe Upstream diverts the stream for some project of development of his own land, then Irving Downstream's water supply from the river will be reduced. How may we represent this as a game in extensive form? It depends on the regime of property rights. Under *riparian rights,* Irving will be able to sue Joe and recover any damages that result from the diversion. (This will imply some legal costs that will have to be borne by one landowner or the other.) On the other hand, if the regime is non-riparian, a landowner is allowed to develop his own land as he chooses, regardless of the results for other landowners up or downstream. Thus, in a non-riparian regime, Irving would not be able to sue to recover damages. If he did file a lawsuit he would lose.

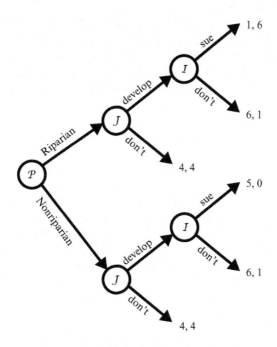

Figure 6.3   Games 6.3, 6.4, 6.5: a public decision process

Let us consider first a case in which the project is inefficient, that is, it imposes costs (damages) on Irving that are greater than the benefits that Joe obtains. Then we have Game 6.3, shown as the upper branch of Figure 6.3, beginning with node *j*. As usual, the payoffs are arbitrary but chosen to be consistent with the assumptions of the problem; in this case total benefits are evaluated on a scale from 0 to 10. The first payoff is to Joe, the second to Irving. The subgame perfect equilibrium of this game is that Joe does not develop his property, and in this instance it is an efficient result; though we should stress that this depends on the particular numbers in this case and if (for example) the benefits of the development were greater, this result could be inefficient.

Now suppose instead that the regime were one of non-riparian rights. In that case Joe has the right to develop his property as he may please, and Irving has no grounds for any lawsuit. If he does sue he will lose. We then have Game 6.4, the subgame shown as the lower branch of Figure 6.3 beginning, again, with node *j*. In this case the subgame perfect equilibrium is that Joe goes forward with the inefficient development, and Irving does not sue.

Riparian rights provide an example of alternative systems of property rights, with no clear rationale other than efficiency for choosing one system over the other. From a normative or ethical point of view, we have two arguments that seem to offset one another: (1) one ought to be able to enjoy an unaltered river frontage; and (2) one ought to be able to develop one's own property as one may choose. In any case the decision between two property rights regimes can only be made by the public authority.

Now let us assume that these two examples are typical ones: that is, that on the average, riparian rights will deter inefficient projects and admit efficient ones (where the damages are less than the benefits to the developer). This is an independent judgment of fact and would have to be verified by empirical research. We suppose that this has been done. We treat the public authority as a player in the game, and assume that the payoff to the public authority is the sum of the payoffs to Irving and Joe. Then the decision of the public authority is represented by Game 6.5, Figure 6.3 as a whole.

The subgame perfect equilibrium of Game 6.5 is that the public authority chooses riparian rights and Joe elects not to proceed with the inefficient project. Note that for this example, both the two-person games with riparian and with nonriparian rights are imbedded games in the public decision. Conversely, in order for noncooperative game theory to be applied validly in the formation of public policy, it is essential that the private-sector decisions be imbedded as subgames in the larger game comprising the public policy decision – that is, that the interdependent decisions in the private sector be analyzed as complete games, and not as fragmentary nested (but not imbedded) games. If we think of public policy as setting "the rules of the game" for the private sector, then clearly private sector decisions must be imbedded in the game of determining public policy.

In this example, the strategies available to Joe and Irving do not have any influence on the decision of the public authority. That is, strategies involving lobbying and bribery have not been taken into account. If these instances of "rent-seeking behavior" are available to the private agents, then their game (that is, the private sector) is no longer imbedded in the public policy decision, although it is nested in that larger game. In that sense, the imbedding of private sector decisions in the larger game of determination of public policy is an ideal case.

6.4   REPEATED PLAY

In a very early experimental study in economics, two experimental subjects played a noncooperative game 100 times in succession. The results did not agree with the predictions of Nash equilibrium for the individual plays. John Nash's response to this was that it was not a valid test of the theory, but that the entire series of 100 repetitions would have to be solved as a single game. (This account follows Poundstone, 1992, Chapter 6.) It was widely suspected that repeated play would make a difference (von Neumann and Morgenstern conjecture along these lines in passing) and could result in cooperative play in an otherwise noncooperative game. This suspicion was so strong that it is sometimes called a "folk theorem," although it has not been proved, nor even formally stated, and indeed is not true in general. However, the tools to deal with the question did not exist until the 1970s. A game consisting of repeated plays of a simpler game is best dealt with as a game in extensive form, and subgame perfect equilibrium is a key tool.

6.4.1   The "Folk Theorem"

Nutter's (1964) theory of oligopoly envisions price competition among a small number of firms as a Nash equilibrium in which the competitive price is the unique equilibrium. This equilibrium can be arrived at by iterated elimination of dominated strategies,[3] in which a lower price always dominates a higher price (at or above the competitive price). However, typically, oligopolists face one another in price competition again and again, period after period. Thus, perhaps they will be able to attain a cooperative solution (among themselves, not considering the customers as players) and share monopoly profits based on higher prices. If so, there would be important implications for antitrust policy. This is one of the most important (and common) applications of the theory of repeated play.

As usual we begin with a simple example, designed to minimize mathematical difficulty. Firms A and B will choose between just two prices, high and low, and the prices are the behavior strategies for the duopolists. The high price corresponds to a shared monopoly and so higher profits for both firms, and the lower price corresponds roughly to a competitive price. We will not explore the details of cost and revenue that generate these profits and, for mathematical simplicity, will ignore intermediate, lower and higher prices that might be charged. The example is Game 6.6, shown as Table 6.1. However, the two firms will play the game again and again, and this repeated play, taken together, has to be analyzed in extensive form.

Table 6.1   Game 6.6: duopoly

Payoffs: A, B		Firm B	
		High	Low
Firm A	High	7,7	3,10
	Low	10,3	4,4

The strategies for this game are behavior strategies, while a (von Neumann and Morgenstern) contingent strategy for this game would be a sequence of conditional decisions for each play of the game.

Case 1   M repetitions

We first consider a case in which the play is repeated for a definite number of iterations, indicating the number of repetitions as M. The repetitions of Game 6.6 are indicated as 1, 2, . . ., M. In the larger game of repeated play, subgames[4] are sequences j, . . ., M, where $1 \leq j \leq M$. In particular, play M, the last play, is the only basic subgame. We know the equilibrium of that subgame: it is "low, low" for payoffs 4, 4. We then consider the reduced game consisting of repetitions 1, . . ., M − 1, with payoffs augmented by 4 each.[5] The only basic subgame of this game is repetition M − 1. Once again, we know the equilibrium and it is the noncooperative solution in behavior strategies to Game 6.6. We continue in this vein until repetition 1 is the reduced game; for it, too, the noncooperative solution is the equilibrium. The conclusion is this: for a game with a definite number of repetitions, the folk theorem is false, and the only subgame perfect equilibrium is a sequence of plays of the Nash equilibrium behavior strategies in the original game.

Case 2   Indefinite repetitions

Aumann (CTG4, 1959, pp. 287–324) defined a *supergame* for a game Γ as an infinite sequence of repetitions of Γ. Clearly, the reasoning in the previous case will not apply to a supergame, since the supergame has no basic subgames. Every subgame is a sequence of repetitions of Γ indexed as j, j+1, . . ., without limit, so every subgame contains other proper subgames. We will have to use different methods to deal with supergames.

But is this realistic? After all, nothing lasts forever! However, Case 1 seems a little artificial in assuming a *definite* number of repetitions. How likely is it that oligopolists, or others engaged in a repeated game, would anticipate the exact number of repetitions that will occur? Suppose instead that at each repetition of Γ, the players can expect that there will be yet another repetition with probability δ, but the probability that there will be no further repetitions whatever is $1 - \delta$. Let y_j be the payoff to a player in the *j*th repetition of the game. Then at repetition *t*, the player wants to maximize the mathematical expectation $\sum_{j=t}^{\infty} \delta y_j$. This formula is the same as a formula for the discounted present value of the series of payments at a discount factor δ, and accordingly δ is referred to as the discount factor.[6] Although the probability of more than M rounds of play approaches zero as M increases without bound, a game such as this has to be analyzed as an infinite game and has no basic subgames.

We may suppose that the players in the game choose behavior strategies for each repetition according to some rule. The "tit-for-tat" rule is an important possibility: begin by playing the cooperative behavior strategy, "high" price in this case, and continue playing it unless the other player plays noncooperatively ("low" price). If the other player plays noncooperatively, then retaliate by playing once noncooperatively on the following round. Notice that the threat of retaliating by playing noncooperatively is credible, since noncooperative play is an equilibrium behavior strategy on any particular round.

Tit-for-tat is called a "trigger strategy," since noncooperation triggers a retaliatory act of noncooperation. However, properly speaking, the tit-for-tat rule is not itself a strategy.[7] It is neither a behavior strategy nor a contingent strategy as understood by von Neumann and Morgenstern. Rather, it characterizes an infinite family of contingent strategies or of sequences of behavior strategies for this game. However, because the retaliation is itself Nash equilibrial, each of the contingent strategies in the family is subgame-perfect, provided that the threat is sufficient to deter the other player from choosing the noncooperative strategy. That is the question to which we now turn.

The question is this: supposing Firm A plays according to the tit-for-tat rule, will firm B be deterred from a *single* opportunistic noncooperative play, that is, from playing "low" at round t, taking advantage of A's cooperative play, and then returning to playing "high, high, high" so long as the game continues. This implies a sequence of payoffs $y_t = 10$, $y_{t+1} = 3$, $y_{t+2} = y_{t+3} = \ldots = 7$. The alternative is to play cooperatively on every round, which implies payoffs $y_t = y_{t+1} = y_{t+2} = y_{t+3} = \ldots = 7$. The expected value of the first sequence is $10 + \delta^3 + \delta^2 7 + \delta^3 7 + \delta^4 7 + \ldots$. The expected value of the second sequence is $7 + \delta 7 + \delta^2 7 + \delta^3 7 + \delta^4 7 + \ldots$. Since only the first two terms differ, the second sequence of payoffs is greater if $7(1 + \delta) > 10 + 3\delta$. A little algebra tells us that this will be true whenever $\delta > 0.75$.

What if Firm B plays noncooperatively again and again? If so, then A will respond by also playing noncooperatively on every turn, in accordance with the tit-for-tat rule. Thus, Firm B's sequence of payoffs is $y_t = 10$, $y_{t+1} = y_{t+2} = y_{t+3} = \ldots = 4$, which can be written as $10 + (\delta/(1 - \delta))4$. For any $\delta > 0.25$, the expected value of this sequence will be less than the expected value for the sequence of payoffs for a single noncooperative play. Conversely, if the threat implicit in tit-for-tat play is sufficient to deter a single round of noncooperative play, it is undoubtedly sufficient to deter systematically noncooperative play. With $\delta > 0.75$, playing against tit-for-tat, Firm B will simply find that more noncooperative means lower expected value payoffs.

What we have found is that, if the probability of another round of play is great enough,[8] in this example, a tit-for-tat rule by one player will make it unprofitable for the other player to deviate from cooperative behavior. If each player plays tit-for-tat, then the play is always cooperative, and neither player can gain anything by deviating from the tit-for-tat rule.

Unfortunately, that is not the whole story. Mutual play of a tit-for-tat rule is only one of many equilibria of an indefinitely repeated social dilemma. In particular, pure noncooperation by both players is always also an equilibrium. There are many others at intermediate levels of efficiency. Nor is the tit-for-tat rule dominant over all other rules by which the game might be played. Suppose, for example, that Firm A plays tit-for-tat while Firm B plays a more "forgiving" trigger strategy rule, tit-for-two-tats. That is, Firm B plays cooperatively unless Firm A plays noncooperatively for two rounds in succession, and then responds with one round of retaliatory noncooperative play. These two rules would lead to cooperation, and Firm B can do no better so long as Firm A sticks to tit-for-tat. But Firm A can do better by deviating from tit-for-tat. In particular, suppose Firm A adopts the rule of alternating cooperative and noncooperative play. Then Firm B never retaliates and Firm A alternates payoffs of 10 and 7, a sequence that dominates the sequence from steady cooperative play. The point is that there are some strategy rules (for example, tit-for-two-tats) against which the tit-for-tat rule does not produce best responses.

The tit-for-tat strategy rule and variants of it, such as a tit-for-two-tats and two-tits for-a-tat (retaliate with two rounds of noncooperative play for one round by the other player) are all *forgiving trigger strategy rules*, which means that the retaliating player will eventually return to cooperative play if the other player does so. A rule that plays cooperatively until the other player initiates noncooperative play and then retaliates by playing noncooperatively on all successive plays is called the *grim trigger*. The grim trigger may deter noncooperative play where tit-for-tat would not. The grim trigger played a key role in warfare in the twentieth century. Poison gas was used as a weapon of war in World War I, and in the Iran–Iraq war of the 1980s, but not in World War II. The use of a weapon such as poison gas may be a social dilemma for the belligerents (McCain, 2014b, pp. 60–61, 360–363). In a long war, with repeated battles, perhaps restraint might be based on fear of retaliation from an opponent playing according to a grim trigger rule. In fact, historical evidence makes it clear that Germany, the United States, and Britain (with pressure from the United States) were following a grim trigger rule with respect to gas (Harris and Paxman, 2002). This example may illustrate the real possibility of cooperation in games of completely opposed interest, but also underscores that there is nothing inevitable about this, and that non-cooperation is always among the equilibria of repeated games.

This discussion assumes a two-person game. The extent to which the results may be extended to games of more than two persons remains a somewhat open question. What is clear is that the relatively simple argument along the lines of the previous example is not applicable to more than two players. Difficulties arise with as few as three players (Fudenberg and Maskin, 1986, p. 543). Allowing for correlated strategies (with public signals) and assuming sufficient diversity in the payoffs to the different players, Fudenberg and Maskin do extend the model to n players. Abreu et. al. (1994) follow Fudenberg and Maskin with a more precise characterization of the conditions for cooperation in n-person games. In a working paper, Haag and Lagunoff (2005) find that diversity in subjective rates of time discounting make cooperation less likely, though it grows more likely in larger groups. Nevertheless, it seems widely felt that larger groups are less likely to cooperate, on the basis of experience in the applications to price competition.

6.4.2 An Extension

In some ways the probabilistic repeated play model seems very plausible. After all, retaliation is a matter of common experience. However, it allows very little for changing circumstances outside the control of the agents in the game. Suppose, for example, that two firms play according to the tit-for-tat rule for a number of years, and then it becomes known that one of them is financially impaired and may go bankrupt. As a result, it seems far less probable that there will be further rounds of "play," and the cooperative agreement breaks down. The repeated play model, with its constant discount factor, does not seem to allow for this sort of possibility. This section will sketch a modest extension of the model that will "realistically" allow for such changes of circumstances to affect the continuation of cooperation.

A key tool for this purpose is the state–transition matrix. We suppose, for example, that there are just three possible states of the world: state 1, in which both firms are financially sound and the "game" of price competition takes place; state 2, in which the "game" takes place but one firm is financially impaired; and state 3, in which there is no play, perhaps because one firm has gone bankrupt. There are two players. In states 1 and 2 they play Game 6.6. In state 3 they do not interact at all, and payoffs for both players are zero.

Given that the world is in state i in period t, the probabilities[9] that the world will be in state j in period $t + 1$ are known constants summarized in the state transition matrix. Suppose the probabilities are as shown in Table 6.2.

Table 6.2   Transition matrix 1

		Transition to		
		1	2	3
Transition	1	0.8	0.2	0
from	2	0.6	0.2	0.2
	3	0.1	0.1	0.8

The number in a given cell tells us the probability that the state represented by the row will be succeeded by the state represented by the column. Thus, for example, this transition matrix tells us that state 1 will be followed by state 1, 80 percent of the time, by state 2, 20 percent of the time, but never directly followed by state 3. Nevertheless, we might see the system in state 1 in the first period, in state 2 in the second period (with 20 percent probability) and in state 3 in the third period. The probability that the system would transit from 1 to state 3 so quickly is the compound probability, $0.2 * 0.2 = 0.04$ – a small probability. But, given more time, the probability could be greater since there are very many more ways that the transition could occur. Using compound probabilities, we can compute the probability that any one of the states will occur in any future period, starting out from state 1 (or indeed any other state). For example, the probability that we will observe state 1 steadily approaches a stable value of 0.64; and similarly for the other states approaches the constants 0.18, 0.18. In fact, many such models have equilibria of this kind, and the equilibria can be found by a fairly simple exercise in linear algebra, solving a system of three equations with the three constant probabilities as the three unknowns. We shall skip the details. We can also compute the probability of yet another round of play, that is, the probability that either state 1 or state 2 will occur in the period n, if play took place in period $n - 1$. (This reflects the probability both that state 1 or 2 will occur in period $n - 1$ and the probabilities in the state transition matrix.) This approaches a constant value of 0.78.

For a case like this, we might just take the equilibrium probabilities and treat the model as if it had constant probabilities, at least as a first approximation. Let us do that, asking whether tit-for-tat play will deter defection in Game 6.6. We find that if the probability of yet another round of play is greater than 0.675, indeed it will. If we begin from state 1, the probability of another round of play is greater than 0.675 in *every single period*, so we can be confident that cooperation is feasible based on the tit-for-tat strategy rule.

Table 6.3   Transition matrix 2

		Transition to		
		1	2	3
Transition	1	0.8	0.2	0
from	2	0.4	0.2	0.4
	3	0	0	1

For our example an advantage of this approach is that state–transition models can represent irreversible events, such as bankruptcy and death. Consider the transition matrix in Table 6.3. Row 2 tells us that when a firm is financially impaired, it will return to financial health with a probability of 0.4, go bankrupt with probability 0.4, or continue impaired in the next period with probability 0.2. As for the third row, it reminds us that liquidation is irreversible: once you are dead you stay dead, and the probability of coming back from the dead is zero.

If we repeat the enumeration of the probabilities of the three states for future periods, beginning in state 1, we see that the probability that state 3 will be observed approaches 1, and the probabilities of the other states, and of another round of play, approach zero. That is, state 3 is what is called an "absorbing state:" sooner or later we are all dead. As a result, the probability of another round of play keeps dropping and approaches 0 in the limit. It may seem that we can apply backward induction so that there will be no cooperation.

However, this is a mistake, or at least hasty. Assume that the agents can observe the state of the world. At the very least, agents will be able to tell whether anyone is bankrupt or not. We will assume that they can also observe whether they are in state 1 or state 2. Therefore, they can make their strategies contingent on the state. Thus, in place of tit-for-tat, suppose both parties play according to Rule 1:

Rule 1. Cooperate IF the state is 1 AND (it is the first round of play OR the state in the previous period was other than state 1 OR the other agent played cooperate on previous round) ELSE defect.

Now suppose we are at state 1 and one player defects on the current round, planning on returning to "cooperate" thereafter. His expected payoff is $10 + 3*0.8 + 4*0.2 = 13.2$. On the other hand if he cooperates the expected payoff is $7 + 7*0.8 + 4*0.2 = 13.4$. Cooperation pays better and defection is deterred. The term $4*0.2$ is the payoff of the mutual defection that is sure to occur in case state 2 is realized in the next period times the probability that this will happen. (The example does not allow for time discounting and with time discounting the result might be different.) In this case the probabilities 0.8 and 0.2 are always applicable because they are conditional probabilities, conditional on the observation that state 1 has occurred.

Suppose instead that state 2 has been realized. Then the conditional probabilities of state 1 and state 2 in the following period are 0.4 and 0.2. Suppose the player defects once while the other player plays Rule 1. Then the defector's expectation is $10 + 7*0.4 + 4*0.2 + 0*0.4 = 13.6$. If he plays cooperate it will be $7 + 7*0.4 + 4*0.2 + 0*0.4 = 10.6$. (Since there is no play in state 3 we assign payoffs of zero.) Cooperation does not pay.

Thus, whatever state occurs, there is no incentive to deviate from Rule 1 – Rule 1 is subgame perfect. (Here, again, we are assuming the rate of time discount is sufficiently small.) But notice what it means. We start from state 1, with cooperation. Over the next few rounds, the probability (as seen from period 1) that we will remain in state 1 declines. We can foresee that within several rounds, with high probability, the system will transit to state 2, and at that point cooperation will break down. If the firm in trouble manages to return to financial health (the system transits back to state 1), cooperation will be resumed. On the other hand, if one firm is liquidated, there will be no more opportunities for cooperation; and since this will occur sooner or later, cooperation will be repeated only a finite number of times.

There could be a range of other applications and contingent rules. For example, the agents might be playing different games in different states, with play in one game contingent on the other's strategies either in the last play of the game now being played, or in the last play of the other game, or both.

6.4.3 Interim Summary

We see that repeated play can be a link from Nash equilibrium to cooperative play. If agents are involved in interactions that are likely to be repeated, and the agents are patient enough and have some foresight, then cooperative play may emerge as one of the equilibria in a supergame, that is, a game repeated for an indefinite number of times.

6.5 ON SOME EXPERIMENTAL STUDIES

A number of experimental studies have addressed the predictions of the perfect equilibrium model in noncooperative game theory. Two games are

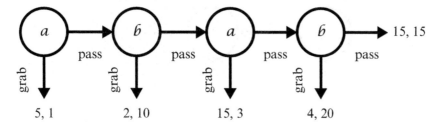

Figure 6.4   Game 6.7: a Centipede Game

particularly important in this connection: the Ultimatum Game and the Centipede Game. On the whole, the experimental results disagree strongly with the predictions of the perfect equilibrium model, if it is considered as an empirical hypothesis. The Ultimatum Game has been discussed in historical context in Chapter 3. Here we will focus on the Centipede Game.

The "Centipede Game" (Rosenthal, 1981; McKelvey and Palfrey, 1992) is illustrated by Figure 6.4. The centipede is a game with two participants and a pot of money payoff dollars. The two participants will be *a* and *b*. The game proceeds in stages, and at each stage one or the other of the participants must make a commitment. At the first stage player *a* can either take or pass a money payment. If he takes it *b* also gets a smaller payment. If *a* passes at the first stage, *b* has an opportunity to take a larger share of the payment, leaving *a* the smaller share. However, if *b* passes at the second stage, *a* in turn gets an opportunity to take the larger share, and the game proceeds in this way. The two players alternate, as shown in Figure 6.4, where the numbers show the payoff to *a* first and then to *b*. The game ends after some finite number of steps with each participant getting a specified share of the pot. The size of the total payment to the two players may increase with the number of stages the game continues. This could be a model of "roundabout" production in economics, in that "passing" the pot on an early round allows the resources generated in the first round to be compounded in the later rounds. In some studies the game has subsequent stages, and it may have many stages. If we visualize a game with 100 stages rather than four, the basis of the name "Centipede Game" becomes clear.

A cooperative solution to this game requires a sequence of behavior strategies "pass." Using backward induction it is clear that the subgame perfect equilibrium in this game is for *a* to "take the money and run." Since *a* knows that *b* is a rational player, *a* cannot expect that *b* will pass on the second round and allow *a* to grab the larger amount, 15, at the third

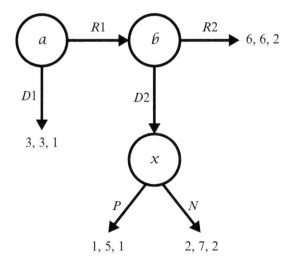

Figure 6.5   Game 6.8, 6.9 in extensive form

stage, nor can *b* expect that if he passes *a* will allow him the opportunity to grab 20 at the fourth stage. In short, a rational agent cannot "outsmart" another rational agent.

The experimental evidence does not agree, and a variety of outcomes (but only rarely the cooperative outcome of continuing play to the end) are observed. Many of the observed sequences of play are consistent with the possibility that one or both players are trying to "outsmart" one another – with at least one of them failing to do so. Suppose that many experimental subjects commit themselves to inconsistent rationalizable strategies such as "If *a* passes then I will pass at the second stage and then, if he passes again, I'll grab at the last stage for 20 rather than 10." If *a* conjectures that *b* has adopted that strategy, *a*'s best response would be "Pass at the first stage and then, if *b* passes, grab at the second for 15 rather than 5." On this interpretation, the evidence suggests that, at least in some circumstances, experimental subjects may adopt inconsistent rationalizable strategies.

In studies of reciprocity, variants on the centipede have given rise to important results. Figure 6.5 will serve as a generic diagram for two extensions of the centipede. These games take place in a maximum of three stages, although it can be cut short by either player.

In Game 6.8, decision node *x* is player *a*'s second decision node. (For now, ignore the third payoff number.) It introduces a "punishment" or "threat strategy," *P*, that gives *a* the option of reducing *b*'s payoff at the cost of some reduction in his own, or of not doing so (arrow *N*). The basic subgame is the punishment node, and its Nash equilibrium is N. The game is then reduced to a two-step centipede with payoffs of 2, 7 at the second stage and $D1$ is the subgame perfect behavior strategy for a. The prediction of the subgame perfect equilibrium model is that the punishment node will make no difference and a will grab at the first opportunity, just as in a Centipede Game without the punishment node. However, such punishment is often observed, and cooperative outcomes (with payoffs such as 6,6) are more common in this game than they are when the third stage does not exist in the experiment. Since a decision for P at the punishment node leaves a worse off than he would be otherwise, P would be an instance of negative reciprocity.

Now suppose instead that the third node in Figure 6.5 is not the decision of player a but of a third player, c. The payoffs to c are shown third. This is an example of third-party punishment. As the choice of P makes agent c worse off, it would be an instance of altruistic punishment. It may also be referred to as "third-party reciprocity" (Fehr and Fischbacher, 2004). Once again, noncooperative game theory predicts that such third-party punishment will never occur, and consequently the strategies of a and b will be the same as they would be if there were no third stage; but the experimental evidence does not confirm this prediction. Rather, third-party punishment is observed, and seems consistent with the hypothesis that the third parties place some value on reciprocity between the original two players, and punish deviations from it (Fehr and Fischbacher, 2004).

On the whole, then, experimental evidence does not favor the subgame perfect equilibrium as a general empirical hypothesis. On the other side, we should observe that these experiments have themselves arisen from the subgame perfect equilibrium model. By drawing on that analysis, they have provided more precise evidence on human motivation than earlier experiments were able to supply. Thus, we may best regard the subgame perfect equilibrium as defining one extreme of a spectrum of forms of rationality that we may observe in human action, a point to which we return in Chapter 9.

6.6   SUMMARY AND CONCLUSION

Games in which play takes place as a sequence of decisions, so that the decision maker at any stage can condition his decisions on previous decisions of others, are often expressed in extensive form. For these games, Nash equilibrium can be refined, limiting the noncooperative equilibria to those in which the play for each subgame is Nash equilibrial or which are the limit of a series of games in which play is perturbed by a "trembling hand." These are "perfect equilibria," and can be found by backward induction. Perfect equilibria are important in many economic applications, including games of market entry and entry limiting strategies, and principal-agent interactions. When we consider private decisions as games imbedded in the larger game of public policy, we rely on perfect equilibria to assure us that the noncooperative behavior of the private agents will itself be equilibrial. Repeated play is also analyzed in terms of perfect equilibria, with the idea that a threat of retaliation will be credible only if it is subgame perfect. On this basis, conclusions can be drawn as to whether cooperative play is likely to emerge from repeated noncooperative play in a particular case. However, experimental results suggest some caution, in that subgame perfect equilibria may not be realized when they conflict with reciprocity motives.

NOTES

1. For further examples, see my introductory textbook, *Game Theory: A Nontechnical Introduction to the Analysis of Strategy, South-Western*, 2014b, especially Chapters 12, 13.
2. In this as in other cases, "principal" and "principle" may be confused, especially since "principal" is more usually an adjective. However, standard dictionaries concur that this is the correct usage for one who authorizes an agent to act on her or his behalf.
3. Note Game 4.2, Chapter 4.
4. In a more complex game such as Game 6.2, the set of subgames will include the repetitions of the original game, but may be a much larger set. The reasoning is slightly more complex in an example of that kind, but the conclusion is not changed.
5. It may be appropriate to discount the payments at this last stage to present value at some specific discount rate. The economic literature on repeated games tends to stress this, but it does not affect the qualitative results and we shall ignore discounting here.
6. If there is a definite time period between repetitions, then δ should reflect the time discount rate as well as the probability of repetition. See McCain (2014b, p. 273) for details.
7. In what follows, I will use the term "trigger strategy rule," accordingly.
8. If payoffs are discounted for time, then a great enough time discount might offset the high probability of repetition, so that the tit-for-tat strategy rule will fail. Thus, many authors express the point this way: if the players are patient enough (have low enough time discount rates), then cooperation can be attained via tit-for-tat play. In this example, though, time discounts are likely to be small relative to probabilities that play will be discontinued that are in the range of 0.25 to 0.5.
9. Note that this discussion differs from Shapley's (1953) stochastic games, in that for Shapley's model the probabilities depend on the strategies chosen, while for this discussion the probabilities are given.

7. Social mechanism design

The 2007 Nobel Memorial Prize in Economics was awarded to Hurwicz, Maskin, and Myerson for their contributions to social mechanism design. Maskin and Myerson are well known as game theorists, and the scientific background document for the prize (Royal Swedish Academy of Sciences) states "By using game theory, mechanism design can go beyond the classical approach," so arguably this was the third Nobel for game theory. In 2012 the prize was awarded to Lloyd Shapley and Alvin Roth for mechanisms that match candidates and opportunities in a way that is in some senses both optimal and stable, and the prize committee refers to a number of cases in which the Shapley–Roth mechanism has been incorporated by actual organizations. These examples illustrate how game theory can be integral to the improvement of organizations and policies.

"Mechanism design, Professor Maskin explained, can be thought of as the 'reverse engineering part of economics.' The starting point, he said, is an outcome that is being sought, like a cleaner environment, a more equitable distribution of income or more technical innovation. Then, he added, one works to design a system that aligns private incentives with public goals" (Lohr, 2007). Much of the work of social mechanism design has been within the scope of game theory, though (like bargaining theory) social mechanism design has a longer history than game theory does. Hurwicz[1] (1973 p. 2) traces the idea to utopian socialism.

Hurwicz's founding lecture was concerned (in its first three sections) with dynamics and with the exchange of information necessary to achieve efficiency or other objectives by particular mechanisms, supposing that people report honestly. Some mechanisms will require more or bigger (thus more costly) messages to attain efficiency, or even feasibility. The feasibility of central economic planning (at that time allegedly practiced in the Soviet Union) was a major concern. In the fourth and last section he addresses *incentive compatibility*, noting that agents may have incentives to lie, so that honesty may not be taken for granted. A mechanism free of the incentive to lie is "incentive compatible." This, he says (Hurwicz, p. 23) "is a problem in the theory of games, in this case noncooperative games without side payments." This has been the focus of most of the subsequent work denoted as mechanism design or implementation theory.

Mechanism design can be thought of in terms of imbedded games. We have a population of $n - 1$ "real" players, human agents with their own intentions, preferences, and potential courses of action. The nth player is the designer, an "artificial player." For the designer, the various strategies are the various "rules of the game" according to which the $n - 1$ "real players" might play their game, such as riparian or non-riparian rights in Game 6.5. Since these various rulesets define games that are imbedded in the designer's game, perfect equilibrium in the designer's game means that each is in a Nash equilibrium. The "artificial player's" payoff may be an index of the efficiency or equity of the perfect equilibrium, or it may be 1 in case the outcome corresponds to a particular cooperative solution and 0 otherwise.

7.1　CUTTING THE CAKE

As an instance of mechanism design or implementation, we can turn to an idea much older than game theory: cutting a cake. The objective is that the cake should be divided equally – or if not equally, then fairly. Suppose the cake is to be divided between two identical persons, each of whom prefers more cake to less. One of the two is assigned to cut the cake, and the other gets to choose which piece he will take. Then the cake-cutter knows that he will get the smaller piece, and thus has an incentive to make the division as equal as possible. Equal division is incentive compatible, that is, consistent with noncooperative decisions by the two recipients of cake. The rules of the game – one cuts and the other chooses – implement the objective of an equal division of cake.

Of course, the devil is in the simplifying assumptions, as usual. Let us make the problem a little more complex. We suppose the cake contains nuts, and the nuts are not randomly distributed – there are more of them on the right (let us say). Now suppose that the two agents are not alike but are of different *types*: agent a, who is to choose, likes nuts and might prefer a smaller piece if it had more nuts, while agent b is indifferent with respect to nuts and just wants a bigger piece, regardless of the quantity of nuts. What is a "fair division" in this case? A division may be fair in this sense if it is non-envious (Foley, 1967): each person gets a piece of cake that he prefers to the piece the other person has, rather than vice versa. Now suppose b is to cut and a is to choose and b knows a's preferences. Then b can cut the cake into two unequal pieces, with more nuts in the smaller piece, such that a will choose the piece he prefers while b is left the piece that he prefers. Thus a fair division (in this sense) is incentive-compatible.

But what if agent b does not know which *type* agent a is – whether agent a likes nuts or is indifferent with respect to them? (Perhaps he even hates them.) b needs this information to know how to cut, for the benefit of both. b can ask a what type he is – but can he trust the answer? Suppose for a moment that a is, like b, indifferent to nuts. By saying that he likes nuts, a would be able to fool b into cutting unequally. a would then choose the bigger piece, and so be better off than he would be if he told the truth. Thus the rules of the game will have to be designed so that each agent will truthfully reveal what type he is (and that is to say, reveal whatever is relevant to the correct decision). This is the *revelation principle*. It may be that (for a particular class of games, or in general) there are no "rules of the game" that will do this, and this issue has been central to mechanism design.

7.2 NASH AND OTHER EQUILIBRIA AS OBJECTIVES OF MECHANISM DESIGN

In the simpler cake-cutting game, we have a noncooperative game and a Nash equilibrium that, at the same time, satisfy the criteria for an equitable division; criteria that in themselves have nothing to do with the noncooperative game. Broadly speaking, this is the objective of social mechanism design. With multiple Nash equilibria, as we have seen, there may be some uncertainty as to whether the agents will find their way to the right refinement of Nash equilibria. The mechanism would be more reliable if, for example, the Nash equilibrium were unique, or if it could be shown that every Nash equilibrium implements the cooperative solution concept or normative objective (Royal Swedish Academy of Sciences; Maskin, 1999). Many studies in mechanism design seek a noncooperative game for which the objective social state is the dominant strategy equilibrium. This would provide a very reliable implementation, since the dominant strategy equilibrium is essentially unique,[2] has a compelling justification in terms of self-interest, and is cognitively relatively easy. However, implementation as dominant strategies is a difficult objective, and it may not be possible even in principle to find a noncooperative game that implements the objective as a dominant strategy equilibrium. This is illustrated by the game theoretic study of elections and voting, an important topic in itself for public policy. Where implementation in dominant strategies is impossible, another common standard is implementation as a Bayesian Nash equilibrium, since such equilibria lend themselves to learning by trial and error. However, multiplicity of equilibria may be a problem in this case, apart from the special case of "Maskin-monotonicity," (Royal Swedish Academy of Sciences; Maskin, 1999), which is a key condition particularly in the study of elections.

7.3　A NEGATIVE RESULT: NONCOOPERATIVE GAME THEORY AND ELECTIONS

Discussion of alternative voting procedures has a long history, with eighteenth-century contributions from de Borda and Condorcet and in the nineteenth century by the author of *Alice in Wonderland*, the mathematician Charles Dodgson. Here is an example. Suppose there are three types of voters on a committee that will vote among alternatives A, B, and C. There are three of type 1, two of type 2, and two of type 3. Their preferences are shown as Table 7.1. Voting will be by majority rule with an agenda: at the first step a choice is made between A and B, and at the second step between the first stage winner and C.

If each voter votes his sincere first preference, A will prevail over B 5–2 and, at the second stage, C will be chosen 4–3. However, this is not a Nash equilibrium. By shifting the first stage votes to B, voters of type 1 can bring about a contest between B and C at the last step, and B will then prevail, leaving type 1 voters better off with B rather than C. Since neither types 2 nor 3 can then improve their outcomes by shifting from sincere first preference voting, this is the Nash equilibrium in this particular voting game. It is an instance of voting manipulation, and, since at least two of the three type 1 voters must shift to voting for B in order for this to work, it is manipulation by a coalition. In this connection, the question for mechanism design is: can we design a better voting mechanism that would be immune to manipulation in this sense? And the general answer is no.

7.3.1　Arrow's Impossibility Theorem

Any account of the study of elections in game theory and economics requires a background account of twentieth-century developments in welfare economics (McCain, 2007). In *The Economics of Welfare*, Pigou had set out a systematic normative economics in terms derived from Mill's rule-utilitarian ethics. Many economists (following Pareto, 1906/1971) felt that Pigou's (1920) welfare economics assumed too much. In particular, Millian utilitarianism assumed that utility is interpersonally comparable

Table 7.1　Preferences for three types

Type	1	2	3	Votes
First	A	C	B	3
Second	B	A	C	2
Third	C	B	A	2

and additive, so that utilitarian values could support (limited) arguments for equalization of income. In place of utilitarianism, the "new welfare economics" of the 1930s and 1940s assumed only that individuals had transitive preferences, but that the preferences could not be expressed in numerical terms that could be compared interpersonally and therefore could not be aggregated as a basis for welfare judgments. However, many studies assumed that preferences could be aggregated to form a social-welfare function, that is, group preferences. This problem was addressed by Arrow (1951), and he proved a negative result: adopting a set of axioms that seemed to express reasonable conditions for a social welfare function, he proved that no function could satisfy them all. Arrow's conditions were (here I follow Satterthwaite, 1975, pp. 203–4):

(1)   rationality, that the social preference is transitive, free of cycles;
(2)   that it is nondictatorial, that is, that there is no individual whose preferences are decisive regardless of the preferences of all;
(3)   independence of irrelevant alternatives, that is, the social preference between two alternatives depends only on the individual private preferences between them;[3]
(4)   citizen sovereignty, that is, that any alternative might be chosen if it is widely enough preferred;
(5)   non-negative responsiveness, that is, that a shift of preferences by which one alternative rises in the preferences of some individuals could not result in its being lower in the social preference.

This "general possibility theorem" led to a great deal of ferment in welfare economics.[4] In particular, it must be that Pigou's maximization of aggregate utility violates one or another of them. In the discussion that followed Arrow's contribution, all of these were questioned, but the independence of irrelevant alternatives was especially a target of critics.

7.3.2   Elections

Arrow's result was not limited to voting systems. Indeed, his result dismissed judgments of the efficiency of markets even in ideal conditions, no less than judgments of the efficiency of elections. Nevertheless, Arrow made extensive use of election-type rules as examples, and it was widely conjectured that similar problems might arise in elections. This conjecture was given formal proof by Gibbard (1973) and Satterthwaite (1975), independently. Their contributions were different in detail, but complementary, and both drew importantly on Arrow's work. Feldman's (1979) very readable exposition is also helpful, especially to the less mathematical reader.

Satterthwaite defined manipulation as a failure to vote in accordance with the individual's own preferences. The person might report a different order of preferences than his own if, due to the balloting procedure, such a report would give rise to a decision he prefers to the one that would follow from honestly reporting his own preferences. If, for a particular voting procedure, this can never happen, then the voting procedure is "strategy-proof." For a strategy-proof voting procedure, "every set of sincere strategies is an equilibrium as defined by Nash" (Satterthwaite, p. 188). But Satterthwaite proves directly that this cannot be the case if the procedure satisfies the Arrow conditions. In fact, Arrow's independence of irrelevant alternatives and non-negative responsiveness together are equivalent to strategy-proofness. Satterthwaite demonstrates that if a voting procedure is strategy-proof then it is dictatorial. Satterthwaite then shows that a rational, citizen–sovereign social welfare function can be derived from any nonmanipulable voting scheme, and conversely, and uses his own theorem to construct a new proof of the Arrow theorem.

Gibbard (1973, p. 589) expresses skepticism about the identification of nonmanipulation with "honest" voting: "Nothing in the structure of a game form tells us which strategy 'honestly' represents any given preference ordering." Accordingly, he characterizes nonmanipulation by an arbitrary function from preferences to votes, and asks whether such a rule of honest voting can be implemented as a dominant strategy. The answer is no, he argues, since a nontrivial voting game cannot have any dominant strategy equilibria whatever – honest or otherwise. Gibbard uses Arrow's result in his proof.

The connection of this "Gibbard–Satterthwaite theorem" to Arrow's result is quite close in a mathematical sense, but substantively it is less close than it is sometimes thought to be. Recall, Arrow's purpose was normative, and the objective was aggregation of preferences. Thus, the last three conditions assert a dependence of social preferences on actual individual preferences. By contrast, Gibbard and Satterthwaite assert (with the same conditions) a dependence of social decisions on preferences as expressed in voting. Their objective is to pose an empirical hypothesis, and one that seems to be true: that voting is always manipulable. The issue of misrepresentation of individual preferences does not arise for Arrow, but, on the other hand, the assumption of nonnegative responsiveness seems much more natural in the normative than the positive framework.

Manipulable voting processes can easily be seen to violate some of the Arrow conditions in simple examples. Return to the example of Table 7.1. Note that the best response of a type 1 voter depends on the fact that voters of type 3 rank C below B. Suppose alternative B is eliminated, so that the choice is strictly between A and C. Type 1 voters can no longer gain by voting strategically and C becomes the winner. This violates the independence of irrelevant alternatives. The example points up a strong connection between the independence of irrelevant alternatives and manipulation of voting.

The negative result in the Gibbard-Satterthwaite analysis is not quite the whole story. Maskin remarks[5] "The Arrow Theorem is too negative" in the context of voting theory. Maskin proposes that elections with more than two alternatives should be conducted by a Condorcet evaluation: (1) voters record their preference orderings, rather than a single choice; (2) all possible pairwise choices are evaluated to determine which would win against the other; (3) if one alternative wins over all the others, it is the choice. But this procedure is not *decisive*. That is, there may be preference profiles such that for at least three alternatives, A is preferred to B, which is in turn preferred to C, and C is preferred to A. Indeed Table 7.1 gives an example of this. Maskin concedes that some other criterion would have to be used as a "tie-breaker" in such a case, but argues that, regardless of the tie-breaker, a procedure that always chooses a Condorcet winner will be more workable than any procedure that does not.

He draws this conclusion despite the Arrow theorem on the following reasoning: the Arrow theorem demands that a particular decision rule satisfy very demanding assumptions for *all possible* profiles of preferences over voters. However, he suggests, a procedure incorporating the Condorcet evaluation could satisfy the same assumptions over *a large proportion* of all possible profiles, while failing only on a subset of the possible profiles. In Maskin's terminology, a decision rule "works well" for a particular subset of preference profiles if it satisfies a list of assumptions[6] including independence of irrelevant alternatives and adding decisiveness for any of those profiles. In Dasgupta and Maskin (2008) Maskin compares simple majority rule (which he understands as the Condorcet evaluation) with plurality rule and the Borda count, a system that awards points depending on the individual's preference ranking and chooses the alternative with the most points. He shows that there are sets of profiles on which each of the three "works well," but that Condorcet evaluation "works well" on the sets in which either of the others work well, while the others fail on some profiles for which Condorcet evaluation "works well." Thus, despite Arrow's negative conclusion, Maskin makes a strong case for group decisions according to the Condorcet evaluation.

There are a number of other important papers in the literature that reconsider one or another aspect of the Arrow–Gibbard–Satterthwaite impossibility theorems, generally in favor of some kind of majority rule. In addition, there is some reason to believe that in elections with a large number of voters, the likelihood of manipulation decreases with an increasing number of voters (for example, Baharad and Neeman, 2002). On the other hand, casual empiricism suggests that even in elections with a large number of voters, manipulation is quite common.

7.4   CAP AND TRADE REGULATION

Tradeable emissions controls have become a popular means of controlling environmental pollution. This sort of program is often called a cap-and-trade program (Colby, 2000, p. 639). This approach does not really arise from the program of social mechanism design and relies less on game theory than on the much older competitive market theory of neoclassical economics (Hahn, 1989, p. 99). Nevertheless, it is similar in its overall objectives and seems to be one of the few ideas that, prior to 2008, commanded a good deal of consensus across the divided American political spectrum (Broder, 2007).

In a "cap-and-trade" regulatory framework, polluters are permitted to emit pollutants up to some limit, or "cap," while the permits may be bought and sold among the different polluters who are regulated. The hope is that competitive markets for the permits will develop, allowing a market price that would indicate the least social cost of reducing pollution by one unit, so that pollution objectives would be met more cheaply and the polluters would face approximately socially optimal incentives to create and adopt less-polluting technologies. (See McCain, 1978, on the influence of market prices on technological trends toward environmental degradation.)

We now have some experience of cap-and-trade programs. In the 1980s Hahn (1989) found some impact on costs, though less than theory would have anticipated, but no significant impact on environmental quality and few trades. To some extent that could be attributed to limits and imperfections in the regulatory programs and lack of competition in the permit markets. Lack of competition in permit markets is particularly likely to be a problem if the objective of policy is to reduce local concentrations of pollutant, since the localities are quite likely to contain only a few large sources of pollution. A decade later, Colby (2000, p. 642) reported that, after a delay of years, trading (in sulphur dioxide emission permits) "took off," and "Twenty years after regulators, utilities, and environmental advocates were introduced to air emissions trading opportunities, . . . cautious exploration has evolved into a mature, productive allowance market." Trade in water access has been less successful (ibid., pp. 643–5). Despite intense resistance, some success is reported in markets for tradable fishing quotas (Colby 2000, p. 647).

This experience indicates that tradable permits may be helpful, but that the form and consequences of such policies are as much a matter of political economy as of neoclassical economics or game theory, that oligopoly in permit markets is often likely to be a problem, and that successful permit programs may require a long learning period and may face crucial obstacles in customary norms, particularly (as in fisheries) where neither regulation nor trading is customary. Considering these difficulties, noncooperative game theory could be useful in taking oligopoly and thin markets into account, but the others are as foreign to noncooperative game theory as they are to neoclassical economics. A discipline of social mechanism design will need to be enriched with elements from cognitive science, political theory, and perhaps cultural anthropology if it is to address these problems.

7.5 THE SHAPLEY–ROTH MATCHING ALGORITHM

In 2012, the Nobel Memorial prize was awarded to Lloyd Shapley of UCLA and Alvin Roth of Harvard for work on game-theoretic analysis and design of matching processes, such as the placement of medical residents and admissions to school programs. Shapley's earlier work, with the late David Gale (1921–2008), provided an algorithm (1962) to match pairs – such as males and females interested in a match for marriage or a date – that would produce matches in the core of the game; that is to say, matches with the property that no two candidates would mutually prefer to be matched with one another rather than remain in the matches they have. In an investigation of procedures of the clearing house matching medical college graduates to internships, Roth (1984) discovered that their procedures were equivalent to the Shapley–Gale algorithm. Since then, school systems have in some cases adopted the Shapley–Gale–Roth approach to match students to schools, with improved results, and other applications and adaptations have taken place.

A key aspect of the algorithm is a "deferred acceptance" procedure. Consider a "game" with two types of agents. A useful coalition comprises one of each type – marriage, for example. They all have different preferences with whom they would prefer to associate. In a multi-stage application process, individuals on one side (proposers) apply to individuals on the other side (responders) in the order of the proposers' preference. At each stage, each responder gives a *tentative* acceptance to the applicant she or he most prefers – but can instead reject that applicant if a more preferable candidate applies in a later round. This will best be understood by means of an example.

For the example, there are four young men who aspire to be apprentice

Table 7.2   Preferences for masters and apprentices

Preference rating	Master carvers' preferences			
	John's preferences	Will's preferences	Abe's preferences	Mark's preferences
1	Ronnie	Jemmy	Jemmy	Jemmy
2	Jemmy	Tommy	Ronnie	Davey
3	Tommy	Ronnie	Davey	Tommy
4	Davey	Davey	Tommy	Ronnie
Preference rating	Apprentice candidates' preferences			
	Davey's preferences	Ronnie's preferences	Jemmy's preferences	Tommy's preferences
1	John	Will	John	Abe
2	Abe	John	Abe	John
3	Will	Mark	Mark	Will
4	Mark	Abe	Will	Mark

carvers and four master carvers, each looking for one apprentice. Their preferences as to the association are shown in Table 7.2.

At each stage, any unmatched candidates submit applications to the masters. At each stage, an applicant can submit only one application, and will apply to the master who is most preferred among those who have not rejected the candidate, in order of the candidates' preferences. At each stage, each Master *tentatively* accepts the applicant he most prefers, and rejects any others. Those candidates who are not matched may submit applications to other masters step by step on subsequent stages. On any subsequent stage, a Master can accept a more preferable candidate and bump the one he has tentatively accepted. At each stage, we suppose that decisions are made simply to obtain the most preferred of the decision maker's alternatives; that is, that no-one misrepresents his "type."

Now let us see how this procedure would go.

(1) The candidates submit applications
   a. John gets applications from Davey and Jemmy and rejects Davey
   b. Will gets Ronnie
   c. Abe gets Tommy
   d. Mark gets nobody
(2) Davey applies to his second choice – Abe
   a. Abe drops Tommy and tentatively accepts Davey
(3) Tommy applies to his second choice, John, and is rejected

Table 7.3   Matches

Matches	
John	Ronnie
Will	Tommy
Abe	Jemmy
Mark	Davey

(4) Tommy applies to his third choice – Will
    b. Will drops Ronnie for Tommy
(5) Ronnie applies to his second choice – John
    c. John drops Jemmy for Ronnie, his first choice
(6) Jemmy applies to his second choice – Abe
    d. Abe drops Davey for Jemmy
(7) Davey applies to his third choice, Will, and is rejected
(8) Davey applies to Mark and is accepted with a sigh of relief

The resulting matches are shown in Table 7.3.

Now let us see how the algorithm has performed. John is matched with Ronnie, his first choice. Ronnie has his second choice but Will, his first choice, is matched with Tommy, whom Will prefers to Ronnie. Will is matched with Tommy, his second choice, but Jemmy, Will's first choice, is matched to Abe, whom Jemmy prefers to Will. Will is Tommy's third choice, but Tommy's higher preferences, Abe and John, are each matched with apprentices they prefer to Tommy. Abe is matched with Jemmy, his first choice. Abe is Jemmy's second choice, but Jemmy's first choice, John, is matched with Ronnie, whom John prefers to Jemmy. Mark and Davey are matched. Davey is Mark's second preference, but his first preference, Jemmy, is matched to Abe, whom he prefers. Mark is Davey's lowest preference, but the other three all are matched with candidates they prefer to Davey. Thus by enumeration, we see that the matches are in the core. They are stable – in that no unmatched pair would be mutually willing to be paired in place of the matches they have – and the matchings are Pareto optimal, meaning that no-one can be made better off without making someone else worse off.

It is clear that this procedure is not symmetrical – proposers and responders make decisions of different kinds. It may seem "unfair" that responders are allowed to change their minds. But this – the "deferred acceptance" – is a key to the advantages of the algorithm. It means that no one has to "second guess" the decision to apply or accept. Suppose, for example, that an acceptance were a commitment for both sides. Guessing (correctly) that he is last in Abe's preferences and that John will be popular, Tommy might instead apply to his third choice, Will, in order to avoid being left with Mark. Will would accept Tommy in preference to Ronnie. John would have to stick with Jemmy and Ronnie would have to stick with Will, even though John and Ronnie would prefer to be matched. Thus the match would be both inefficient and outside the core of the game. It is unstable in that sense. This failure occurs because some participants, Tommy in this instance, may have incentives to falsify their preferences, just as they do in strategic voting in elections. Because the masters are allowed to change their mind – because of deferred acceptance – all can be better off.

This result is robust in experimental studies, applicable, and useful. We may hope that there is the potential for many more discoveries or inventions of this kind in the field of social mechanism design.

7.6 ASSESSMENT OF MECHANISM DESIGN

Marx (1845) wrote, "Philosophers have hitherto only interpreted the world in various ways; the point is to change it."[7] That is the standard by which mechanism design invites evaluation. While a number of designed mechanisms, such as the Groves–Ledyard mechanism for public goods supply (Groves and Ledyard, 1977) remain untried in the real world, the long history of elections provides some experience on the reliability of mechanism design. The universally acknowledged success story for mechanism design, however, is the design of auction mechanisms (Royal Swedish Academy of Sciences; Lohr, 2007; Glenn, 2007; McMillan et al., 1997). Accordingly, this seems to be the appropriate testbed for the theory of mechanism design.

Of course auction theory, like bargaining theory and voting theory, has a history before game theory or the literature on mechanism design. In particular, though, the work of Vickrey (1961) gave rise to the "Vickrey auction," which fits the description of mechanism design. The success of eBay® can partly be attributed to its adaptation of Vickrey's ideas. Vickrey suggested a sealed-bid, second price auction as having desirable properties of efficiency and seller revenue maximization, though (in an expected value sense) several major auction designs could be seen to be equivalent. But Vickrey auctions have their desirable properties in the case of individual private value auctions, that is, auctions (such as those for collectibles) in which individual values are subjective, uncorrelated, and known only to the individuals themselves. For auctions in which the items sold have an objective value, imperfectly known to the bidders, inefficient overbidding might occur – in effect, the auction being won by the bidder who most overestimates the value of the item being sold. This is known as the "winner's curse" (Camerer, 1987). Otherwise, when more than one item is sold and the items sold might be complements or substitutes, sealed bid auctions could be less efficient than ascending auctions, since the latter would provide more flexibility to coordinate purchases. In general, of course, if one buyer were large enough to exercise monopsony power, or if there were collusive (cooperative) behavior among the bidders, Vickrey's results might not apply.

If we consider auctions of resources or privileges that would enter into production or marketing, such as the privileges of drilling for petroleum on government-held territory or of using electromagnetic spectrum, it is unlikely that the values will be subjective and uncorrelated; and it was the auctioning of electromagnetic spectrum that put mechanism design "on the map." As early as the 1980s, New Zealand had begun to allocate electromagnetic spectrum by auction, using a Vickrey auction design. However, some of these auctions had disappointingly small yields as buyer monopoly power and/or collusion limited the bidding. In 1993, three well-known economists associated with the consulting firm MDI Associates invented the simultaneous ascending auction, according to the company's own history (Market Design, Inc., 2007). The simultaneous ascending auction was intended to avoid the "winners' curse" and facilitate adjustment of bids to complementarities among the items being sold, by learning from one another's bidding. This auction design was adopted by the Federal Communications Commission for the allocation of licenses to use electromagnetic spectrum in the United States. This series of auctions was considered very successful and largely gave rise to the conception of mechanism design as a practically successful field.

In 1994–97, there were 13 auctions based on the simultaneous ascending auction. It seems clear that they were indeed successful on the whole, and two were quite remarkably successful (Cramton, 1997). However, a series of European auctions of third-generation telecommunications spectrum licenses were less successful, on the whole, and Klemperer (2002) put the blame on mechanism design, saying (p. 3) "Good auction design is really good undergraduate industrial organization; the two issues that really matter are attracting entry and preventing collusion," and adding in a footnote, "By contrast, a graduate knowledge of modern auction theory is at best of lesser importance and at worst distracting from the main concerns." Klemperer observes that ascending auctions permit collusion for the same reason they avoid the winner's curse: they allow bidders to learn from the bids of others, and this learning may facilitate collusion. Thus, he argues for sealed-bid auctions or auctions with a sealed-bid stage although "Auction design is not 'one size fits all.'" (pp. 3–4). Among the European auctions, all but one used an ascending-auction design.

Of those, the first, the English, was a major success, but by contrast the Netherlands, Italian, and especially Swiss auctions, that followed, could be characterized as failures. German and Austrian auctions differed somewhat from the others, and of those, the Austrian auction too was characterized as a failure. The one exception, Denmark, used a sealed-bid auction design and was successful.

To some extent the European failures were the results of bad timing. They took place during or after the dot-com collapse of 2001. However Klemperer's discussion makes it pretty clear that they failed also because of neglect of the "two issues that really matter." Certainly collusion is a cooperative phenomenon: as we recall, price competition is an exceptional case in which cooperative behavior (among those on one side of the market) is undesirable. The other issue that really matters, entry, is somewhat complex in itself. Klemperer seems to mean two things: first, entry into the auction itself, and second, entry into the market. On the one hand, several of the European auctions failed in part because so few of the potential competitors actually participated in the auction. Broadly speaking, in assembling a group to participate in the auction, the governments who conducted the auctions were assembling a cooperative coalition to act on a common strategy, although the common strategy itself demanded noncooperative action from the bidders (McCain, 2014a, pp. 65–66). On the other hand, for a firm to enter an industry is to seek to form cooperative coalitions (for exchange) with potential buyers. Thus, it seems fair to say that the failed European auctions failed because auction design was narrowly based on noncooperative game theory and neglected cooperative–game perspectives.

On the other hand, the relatively successful British and Danish auctions incorporated Klemperer's concerns, and the US auctions clearly were characterized by predominantly competitive bidding (Cramton, 1997). The American market, larger than those of the European countries, was (on Klemperer's reasoning) more favorable to competitive bidding. Moreover, the two auctions that were most successful had special circumstances promoting competition that were not predicted. In the second auction, of narrowband licenses, special arrangements to encourage minority and female participation led to vigorous bidding by those groups that spilled over into the bidding of others (Cramton, 1997, pp. 433–4, 451). In the December 1994 broadband auction the participation of Craig McCaw, seemingly seeking an opportunity to re-enter the telecommunications market after selling his company, was a highly competitive factor (Cramton, 1997, pp. 454–5). All in all, it seems that Plott (1997, p. 637) goes a bit far when he writes "The overall success of the auctions must be attributed to . . . economic theorists, applied economists, FCC lawyers, and the FCC staff." Or if we accept this assessment, then the failures of some of the European auctions must be similarly attributed.

On Marx's criterion, then, mechanism design must be seen as partially successful. It has indeed changed the world. However, the change in the world has not entirely been the one intended: mechanism design has not quite "made history consciously." Yet this is hardly a damning conclusion. Errors occur in all human activities, and the key to progress (from a pragmatic point of view) is that they are corrected. An improved noncooperative game theory, using alternatives to the Nash equilibrium or selectively incorporating ideas from cooperative game theory and other disciplines, may yet provide public policy with a reliable tool of prescription that can be used to specify policies that could advance a wide range of policy priorities. There is some literature in implementation theory that points in that direction (for example, Moulin and Peleg, 1982). Nevertheless, it seems fair to say that noncooperative game theory has not yet become that tool.

7.7 CHAPTER SUMMARY

In mechanism design, game theory is turned from diagnostic to prescriptive use. This demands more of a theory than diagnostic application does. Noncooperative game theory assumes that human decisions are always noncooperative: and there is little doubt that they sometimes are, but it is at least possible that they are also sometimes cooperative. If this is so, then problems that would arise in a hypothetical noncooperative world may also arise in the actual world. In the actual world, these problems may be less pronounced, or it may be that noncooperative behavior even on the part of a minority of the population will result in a noncooperative outcome, in particular cases. On the other hand, a mechanism design based on assumptions of noncooperative behavior may be undermined when cooperative coalitions are formed to exploit it, as in the case of collusive pricing and bidding strategies. Thus, noncooperative game theory may be far more reliable as a diagnostic tool than it is as a prescriptive tool, and the experience of auction theory seems to support this conjecture. Nevertheless, the experience of auction and market design, matching algorithms and cap-and-trade regulation suggest that the mechanism design perspective can improve (if not perfect) policy making and that there is the potential for sweeping new discoveries in this field.

NOTES

1. Hurwicz's paper is sometimes dated as 1972. It was presented as the Richard Ely lecture at the annual conference of the American Economic Association in 1972, and published in the proceedings volume. At that time, the American Economic Association met in late December of each year and the proceedings volume was issued in the following May. Thus, 1972 is the correct year for the lecture but 1973 is the year for the print publication.
2. If more than one strategy sets are dominant strategy equilibria, all will have the same payoffs.
3. This assumption is stated in different ways, for mathematical analysis, by different authors. Following Dasgupta and Maskin (2008), if alternative A is chosen from the set of available choices X, and if X′ is a subset of X, A is an element of X′, then A is chosen from X′. In other terms, dropping alternatives that are not chosen out of the set of alternatives does not change the alternative chosen.
4. Important further developments arose from the work of Foley (1967), who demonstrated that the ordinal preference theory could be used to make distributive judgments. Indeed, the distributive norms arising from models like Foley's tend to be more equalitarian than the utilitarian ones are, and parallel the ideas of the philosopher John Rawls (1971). John C. Harsanyi (1975) drew on Rawls's ideas but also on the reformulation of utility theory in the von Neumann–Morgenstern tradition and on Bayesian concepts of rationality, and argued that social welfare could after all be based on a summation of individual utilities. Amartya Sen (1969) has proposed conditions less limiting than Arrow's that allow the possibility of a consistent majoritarian social welfare function. Sen, however, rejects what he describes as the welfarism of both the old and the new welfare economics, by which he means the supposition that the goodness of a social system depends only on the welfares of individuals in those social systems. In addition, Sen would have data on the capacities and perhaps freedoms of individuals reflected in the normative evaluation of economic society.
5. This quotation is from memory, from an oral address at Haverford College, Haverford, PA, Nov. 16, 2007.
6. While the list differs somewhat from 1–5 above, it is similar in conception. The literature includes a number of different equivalent and near-equivalent axioms expressing the desiderata of good collective decision processes.
7. One finds a considerable variety of versions of this quotation, partly, no doubt, due to alternative translations from the German.

8. Superadditive games in coalition function form

This chapter reviews some concepts from what might be called near-consensus cooperative game theory. The objective of the chapter is primarily expositional. Apart from expression, examples, arrangement, and some critical comments, the chapter is not intended to be original.

For this chapter, the game is primarily represented in coalition or characteristic function form. That is, the game comprises a set n of players, $a_1, \ldots, a_n$; $a_i \in n$; the enumeration of all subsets of that set, the potential coalitions, and a mapping from subsets to real numbers, the characteristic or coalition function, which gives us the value attainable by each coalition. The value of a coalition C will be denoted as $v(C)$ or $v\{a_1, a_2, \ldots\}$ where $a_1, a_2, \ldots$ are the members of coalition C. As we recall, von Neumann and Morgenstern identified this with the assurance value. The key point is that the value of a coalition is well-defined and depends only on the membership of the coalition. We also adopt the assumption, from von Neumann and Morgenstern, that the game in coalition function form is superadditive; that is, that the value of a merged coalition is no less than the sum of the values of the merged coalitions acting separately.

8.1 SOLUTION CONCEPTS

For a superadditive game in coalition function form, the only rational arrangement is the grand coalition. If the grand coalition is formed, nothing can be lost (since the grand coalition must have a value no less than those of any proper coalitions into which it can be decomposed) and something will usually be gained. All that remains is to determine how the value of the grand coalition will be divided among the decision makers. As we recall from Chapter 3, there are several such solution concepts.

8.1.1 The Core and Related Concepts

Probably the most widely discussed solution concept for games in coalition function form is the *core*. The simple idea behind the core of a cooperative game is that no group can be denied the value that they could obtain if they were to form a coalition and act independently of the rest. As an illustration, consider Game 2.5. A singleton coalition that would produce the public good would then have a value of at most 4, so will not produce the public good. The singleton coalition would face a unified opposition, a two-person coalition, capable of producing two units of the public good. The opposition coalition would refuse to produce the public good, presumably in order to increase its bargaining power, since producing the public good would raise the value of the singleton to 7 or 9. Therefore, $v\{a\} = v\{b\} = v\{c\} = 5$. A doubleton coalition that would produce the two units of public good would be worth 12, whereas if it does not produce its value is 10. Moreover there is nothing the opposition singleton coalition can do to reduce the doubleton's payoff below 12, so the doubleton will choose to produce the public good[1] and $v\{a,b\} = v\{b,c\} = v\{a,c\} = 12$. The grand coalition of all three agents will be worth 24 if it produces three units of the public good and less if not, so it will produce them and $v\{a, b, c\} = 24$.

The payoff to agent j, after side payments are made, is denoted by x_j. A set of payments x_j for the n players in the game, consistent with the value of the grand coalition, is called an *imputation*.[2] Accordingly, suppose $x_a = 5$, $x_b = 5$, $x_c = 10$. Then x_a and x_b can instead form a doubleton coalition and earn 12, which they can divide among themselves. Thus we exclude the imputation 5,5,10 from the core. This process of coalition-shopping is called *recontracting*. In general, by the same reasoning, we exclude from the core any schedule of payoffs that does not satisfy:

1.1. $x_a \geq 5$
1.2. $x_b \geq 5$
1.3. $x_c \geq 5$

1.4. $x_a + x_b \geq 12$
1.5. $x_a + x_c \geq 12$
1.6. $x_b + x_c \geq 12$

1.7. $x_a + x_b + x_c \leq 24$

Adding inequalities 1.4–1.6 we obtain $2(x_a + x_b + x_c) \geq 36$, that is, $x_a + x_b + x_c \geq 18$. Comparing this with inequality 1.7, we see that there are infinitely many imputations that satisfy the criteria for the core in this example. In particular 8,8,8; 8,6,10; and 12,6,6 all are members of the core.

Here is another example, Game 8.1. Once again it will be a three-person game and all singleton coalitions are worth 5 if no production takes place. There are two techniques of production, both of which have economies of scale so that they can be undertaken only by coalitions with two or more members.[3] Technology 1 generates profits of 4 for those who undertake it but produces a polluting waste that has to be assigned to some individual agent (who need not be a member of the group that undertakes production with Technology 1) and reduces that person's payoff by 5. Technology 2 generates a profit of 3 and no waste.

A singleton will face a united opposition that can reduce the singleton's value to zero by producing with technology 1 and assigning the waste to the singleton. Therefore, $v\{a\} = v\{b\} = v\{c\} = 0$. A doubleton can achieve a value of 14 by producing using technology 1, and there is nothing the opposing singleton can do to reduce the doubleton's payoff below 14. Therefore, $v\{a,b\} = v\{b,c\} = v\{a,c\} = 14$. The grand coalition has a value with no production of 15, with Technology 1 of 14, and with Technology 2 of 18. Therefore Technology 2 will be used and $v\{a,b,c\} = 18$.

For this game, in order to prevent any two-person group from dropping out and shifting to Technology 1, $x_a + x_b + x_c \geq 21$ is necessary. Since, however, $x_a + x_b + x_c \leq 18$ is also necessary, there are no imputations that satisfy the criteria for the core of the game. That is, the core for this game comprises the null set. This is often expressed by saying "the core does not exist," but strictly speaking, the core always exists, although (as in this case) it may be null.

Both of these games are symmetrical, but this need not be so. Of course, nonsymmetrical games, in which coalition values depend on the individual members of the coalitions, will be more complex, and in some cases very much so.

There are a number of properties that we might like a solution to have. Two of the most important are that it should never be null and should correspond to a unique imputation. Clearly the core satisfies neither of these. However, there are a number of desirable properties that it does have.

Suppose we have two games played by the same set of players, $\Gamma = (n, v(C))$ and $\Theta = (n, w(C))$, and there are constants α and β such that for any coalition C $w(C) = \alpha + \beta v(C)$. The two games are said to be *strategically equivalent*. Suppose then that whenever **x** is a solution of Γ, $\alpha + \beta\mathbf{x}$ is a solution of Θ. Then the solution is *covariant under strategic equivalence*. In more ordinary terms, it says that the solution will be unchanged by a change in the scale of measurements of payoffs, and that is persuasively a good property for a solution to have. The core has this property[4] (Peleg and Sudholter, 2003, p. 25).

The core also has a property of anonymity, which means that the solution does not depend on the identities of the players except so far as their contributions to the values of coalitions are concerned. To see how this might fail, consider a toy solution concept, which is meant only as a bad example. Assign $x_a = v\{a\}$, $x_b = v\{a,b\} - v\{a\}$, $x_c = v\{a,b,c\} - v\{a,b\}$, and so on if there are more than three players. Now consider another game, Θ, that is a permutation of Γ; that is, we simply take a, b, and c in a different order, such as c, b, a. Nevertheless $w\{a, b\} = v\{a, b\}$ and so on. But if we apply the toy solution concept using the new order we have $x_c = v\{c\}$, $x_b = v\{a,c\} - v\{c\}$, $x_a = v\{a,b,c\} - v\{a,c\}$. For the toy solution concept, the solution depends on how the players are ordered. A solution concept that is independent of this ordering is *symmetrical* and if it is similarly independent of any identity the agents may have apart from what is expressed in the coalition function, it is *anonymous*. The core has these properties (Peleg and Sudholter, 2003, p. 26).

If a solution concept gives each agent a payoff no less than he could get acting as a singleton coalition, it is said to have *individual rationality*. The core has this property (ibid., p. 27).

Thus, despite its shortcomings, the core has some properties that we do want to find in a solution, and it captures the idea that a group of players will in general get at least what they can obtain by acting independently.

8.1.2   Arbitrational Concepts

In 1950–53 two solution concepts for cooperative games were proposed, both of which (unlike the core) provide unique solutions that are never null. These were due, respectively, to Nash and Shapley. Both were derived from systems of axioms that describe properties of a solution that might be considered reasonable or appropriate. Luce and Raiffa (1956) suggested that they might be interpreted as frameworks for arbitration, in that an arbiter would consider the properties of the decision in deciding on the distribution of payoffs among the group. (For more detail see McCain 2013, Chapters 1, 2; 2014b, Chapter 17 section 5.)

8.1.2.1   Nash bargaining
Nash begins by assuming that the payoffs to two interdependent decision makers must fall within a feasible set that is convex and compact.[5] These properties assure that the assignment of payoffs to the two persons will be unique and are trivially satisfied for games in coalition function form. (Nash did not assume transferable utility). He assumes that the decision will have the properties of

(1)   Individual rationality, that is, each agent receives at least as much as he could obtain if there is no agreement.

(2) Pareto-optimality, that is, the decision cannot be improved on for both decision makers simultaneously.
(3) Independence of irrelevant alternatives (see Chapter 7, section 7.2.1).
(4) Solutions are covariant under strategic equivalence.
(5) Symmetry. That is, if the two bargainers are interchanged, the chosen payoffs are interchanged accordingly.[6]

Given these assumptions Nash shows that the vector of chosen payoffs can be computed as the solution to a maximum problem. Let x^*, y^* be the payoffs to the two bargainers if there is no agreement and x, y be their payoffs from the agreement. Then x and y will be chosen so that the product $(x - x^*)(y - y^*)$ is maximized among all the feasible pairs (x,y). This solution assigns unique payoffs to the bargainers and is never null. However, it directly applies only to two-person games and makes no allowance for differences in bargaining power. There are several proposals of extensions to more than two players, but none seems widely accepted (for example Harsanyi, 1956, 1963; Roth, 1979; McCain, 2013, Chapters 2, 6).

8.1.2.2 Shapley value
Shapley proposed a solution concept that would associate a payoff with each of n agents in a coalition function game, where n could be any positive integer. Shapley first adopts a series of three axiomata (CGT, p. 71) that are regarded as necessary characteristics of a solution. In ordinary language, these are that (1) nothing depends on the identity of a player, as distinct from the role (dealer, maker of the opening lead play, etc.) that she plays in the game, (2) the payoffs add up to the total payoff of the grand coalition in the game, and (3) if the same players play two different games, the values in the merged game are the sum of the values in the two games. He then shows (1997, pp. 71–4) that these conditions are uniquely satisfied by an algebraic, permutational formula,

$$\phi_i(v) = \sum_{\substack{S \subseteq N \\ i \in S}} \gamma_n(s)\,(v(S) - v(S - \{i\})) \tag{8.1}$$

where $\phi_i(v)$ is the value assigned to player i in the game characterized by v, and $\gamma_n(s)$ is

$$\gamma_n(s) = \frac{(s-1)!(n-1)!}{n!} \tag{8.2}$$

where s is the number of players in coalition S, n the total number of players in the game, and ! denotes the factorial of the number. Thus, in ordinary terms, the value will be the weighted sum of the individual's contributions to the value of a coalition, for all coalitions in which he might participate. The weights do not lend themselves to a simple ordinary language explanation in themselves. However, Shapley also offers (CGT, pp. 78–9) what he describes as a bargaining process that would generate the value. He writes

> The players . . . agree to play the game v in the grand coalition, formed in the following way: 1. Starting with a single player, the coalition adds one player at a time until every player is admitted. 2. The order in which the players are to join is determined by chance, *with all arrangements equally probable.* 3. Each player, on his admission, demands and is promised the amount his adherence contributes to the value of the coalition . . .. The expectations under this scheme are easily worked out. (emphasis added)

This is as equation (8.1) above, where $\gamma_n(s)$ is the probability that the corresponding payment will be the one offered. It may be that Shapley was following the example of Nash (1953) in this bargaining interpretation.

The Shapley value has a number of desirable properties:

(1) It is Pareto-optimal.
(2) It is covariant under strategic equivalence.
(3) While it is not necessarily anonymous, the Shapley value has a property of symmetry: if the players are permuted, without any other change in the game, their Shapley values are unchanged. (Note that this is not true for the "toy" solution concept of section 1 above. The difference arises from the fact that the "toy" assumes a particular ordering of the agents, while the Shapley value averages over all such orders.)
(4) It has an equal treatment property: suppose that two agents make the same contribution to all coalitions; then they are assigned the same Shapley value.
(5) It is additive (this is the assumption in the introductory paragraph of this section).
(6) It has a null player property: if an agent adds nothing to any coalition then his Shapley value is zero.[7]

The additivity property is crucial and is a defining property of the Shapley value. It may also be the least compelling to intuition (McCain, 2013, pp. 12–13).

8.1.3 Nucleolous

The core and the Shapley value seem to be the most widely discussed and applied solution concepts in cooperative game theory. Of the several other concepts, this book will discuss only the nucleolus. Like the Shapley value, the nucleolus is never null and assigns a unique net payoff to every agent in the game. The nucleolus has the further property that if the core is not null, the nucleolus is an element of it. Thus, the nucleolus can be thought of as a core assignment algorithm; that is, as a basis for singling out a particular imputation in the core when the core is not unique. The nucleolus can be computed by linear programming, although there are some complications, even for a game as simple as this one. The nucleolus has some desirable properties. Like the Shapley value and the core, it is covariant over strategic equivalence and has an equal treatment property. Like the core, the nucleolus is anonymous. It is not additive and lacks the null player property.[8] For more detail on the nucleolus and an extension that will be applied in Part II, see McCain (2013, Chapters 1, 6).

8.1.4 Interpretations of the Solution Concepts

The solution concepts are usually presented as mathematical forms, with their interpretation largely left open. Indeed, they may be susceptible of at least two interpretations, and it may be that more than one solution concept might be adopted for different interpretations.

8.1.4.1 Stability interpretation

Returning to Game 2.5, we have said that imputation $x_a = 5$, $x_b = 5$, $x_c = 10$ should be excluded, since x_a and x_b could then secede and earn 12. But what are we to make of this argument? Critics of the core concept have questioned this criterion along the following lines: if a and b are members of the grand coalition, then they have committed themselves to it, and to some prearranged imputation. For them to opportunistically abandon the coalition to increase their payoffs as a group is then seen as inconsistent for a cooperative-game analysis. The cooperative game solution should represent a binding contract. This *binding contract interpretation* might support Nash bargaining, the Shapley value, or the nucleolus, against the core and related concepts.

Indeed, we might say that the core concept is based on coalitional egoism. But isn't coalitional egoism something we are likely to see in the real world, sometimes?

One possible response to the criticism is that commitments, however binding, are not forever. Thus, even if a and b were to remain with the grand coalition for a time, eventually their commitments would expire and they would be likely to make other arrangements. On this sort of interpretation, the core is a concept of stability. The empirical prediction would then be that imputations outside the core are unstable, and therefore *relatively* unlikely to be observed, since they will be short-lived in case they do occur.

8.1.4.2   Rhetorical interpretation

Another possible interpretation is that the abandonment of the grand coalition by $\{a,b\}$ might not literally take place, but might be a threat made in the course of bargaining over the division of the value created by the grand coalition. This interpretation has been primarily associated with other solution concepts, such as Aumann and Maschler's (1964) bargaining sets (which will be beyond the scope of this chapter) and the nucleolus. Both Nash and Shapley referred to bargaining power as motivation for their models. But the rhetorical interpretation could be applied to the core as well. The empirical prediction then would be that imputations outside the core would be rarely or never observed, since they would be rejected in the bargaining that precedes the formation of coalitions.

8.2   THE PROBLEM OF APPLICABILITY

By comparison with noncooperative game theory, at least, there have been relatively few applications of cooperative game theory. In chapters to follow this book will argue that the simplifying assumptions lead to a theory that is simply too abstract to be useful in a wide range of applications. However, there are a few important applications in economics. We will review three: games of exchange, games of production, and applications to the allocation of cost in a multidivisional organization. The first two are applications of the core, while the last begins from the Shapley value.

8.2.1   The Market as Implementation of the Core

Among the most important applications of cooperative game theory is the study of games of exchange. In a game of exchange, there are two or more types of players, who differ in their endowments of particular goods and services or in their preferences or both. Coalitions are formed for the purpose of making reciprocal transfers of the goods and services, that is, exchanges. The benefit of doing so is that each agent may at the end find himself with a collection of goods and services that he likes better than the endowment he began with.

Most contributions to this literature do not assume transferable utility, but instead adopt the non-transferable-utility approach due to Shapley and Shubik (1952). For simplicity, we will consider an example of trade in indivisible units of two goods.[9] Suppose, then, that we have two traders interested in exchanging olive oil for wine. (This will be game 8.2.) Traders of type a are endowed, at the beginning of the game, with three

Table 8.1   Preferences for a game of exchange

Barrels of oil	Barrels of wine	A's preferences	B's preferences
0	0	16	16
0	1	15	15
0	2	14	14
0	3	12	12
1	0	13	13
1	1	11	11
1	2	7	8
1	3	5	5
2	0	10	10
2	1	8	7
2	2	6	4
2	3	3	2
3	0	9	9
3	1	4	6
3	2	2	3
3	3	1	1

barrels of olive oil, while traders of type b are initially endowed with three barrels of wine, and the barrels cannot be divided. With just two traders, then, an individual may find himself with 0, 1, 2, or 3 barrels of oil and 0, 1, 2, 3 barrels of wine. There are 16 such combinations and, for the purposes of our example, we need to know the preferences of both players with respect to all 16. These are shown in Table 8.1. The preferences are expressed as first, second, and so on, so smaller numbers are better. Thus, for example, the third column tells us that a player of type a prefers one barrel of oil and three of wine (fifth preference) to two of oil and one of wine (eighth preference), while the last column tells us that a player of type b prefers one of oil and three of wine to one of oil and two of wine.

Suppose, then, that it is proposed to exchange two barrels of oil for one of wine. This would please the type b trader, raising him from his twelfth to his fourth preference, but it would reduce the type a trader from his ninth to his eleventh preference; so the type a trader would veto the trade. Suppose, on the other hand, that it is proposed to trade one barrel of oil for two of wine. This would raise the type a trader to his sixth preference and the type b trader to his eleventh. Thus we may say that the allocation[10] that results from the one-oil-for-two-wine trade, two oil and two wine for type a and one oil and one wine for a type b, *dominates* the initial allocation. In general, an allocation **x** will dominate an allocation **y** if there is a coalition at least one member of which is better off, and none worse off, with **x** than with **y**. With only two traders in the game, we need consider only the grand coalition and the singleton coalitions.

As before, the core will consist of all allocations that are undominated. As before, the core may not be unique. For this game, with just one trader of each type, we have

2, 2 for a and 1,1 for b resulting from a 1-for-2 trade
1, 2 for a and 2,1 for b resulting from a 2-for-2 trade
1, 3 for a and 2,0 for b resulting from a 2-for-3 trade

In each of these cases, neither person can do better as a singleton, that is, if no exchange takes place. Thinking in terms of prices, or rates of exchange, we see that the price of a barrel of wine can vary from ½ a barrel of oil to 1 barrel of oil within the core. (At a one-to-one exchange rate, the exchange of one for one is dominated by the exchange of two for two, as an exchange of one each leaves each with his eighth preference while the two-for-two exchange leaves each at his seventh. Put otherwise, at a price of 1, each person will offer two units for trade, and this is the market equilibrium).

Now suppose that we have two traders of each type, and suppose that the allocation proposed is that type a's get 2,2 and b's 1,1, corresponding to a price of ½. Instead, consider a coalition of one a and two b's, and of the b's, suppose b_1 transfers two of wine to a, b_2 transfers one, and a transfers one of oil to each. This leaves a with 1,3, his fifth preference; b_1 with 1,1, his eleventh, and b_2 with 1,2, his eighth. a and b_2 are better off than they were in the proposed allocation, and b_1 is no worse off. (If we allowed wine to be divided, b_2 could offer b_1 a cup or two from b_2's second barrel, to make it worthwhile for b_1 to join the coalition.) Thus, the three-person coalition permits an allocation that dominates the proposed 1-for-2 allocation, and the 1-for-2 allocation (and the price of ½) is no longer in the core. What we see is that with more traders, we may have more complex coalitions, and these impose more constraints on the allocations that can belong to the core, so that the core is smaller in a larger game.

Now, suppose that we have three agents of each type, and the proposed allocation gives a's 1,3 and b's 2,0, as in the 2-for-3 exchange and the price of 2/3. Suppose instead that a coalition is formed of 2 a's and 3 b's with the nine barrels of each type allocated so that each a gets 2,1; two b's get 1,2, and the third b gets 1,3. This means that the a's are at their fourth preference, rather than their fifth as in the proposed allocation, and two b's are at their eighth and one at his fifth preference rather than their tenth, as in the proposed allocation. We now have a 2a and 3b coalition dominating the 2-for-3 exchange, and the price of 2/3 is no longer in the core. Once again, the larger game allows for more complex coalitions that impose more constraints on the allocations in the core, resulting in a smaller core.

In the literature on market games, as in this example, the outcomes are vectors of quantities of different goods and coalitions are formed for reallocation of the initial endowments of goods. Usually goods and services are assumed to be divisible and the conventional neoclassical assumptions are made: the individual may be indifferent between two vectors of goods and services, but preferences are convex, meaning that a weighted average of two vectors of goods and services will be preferred to either of the two vectors that are averaged.[11] It is then demonstrated that

(1) the core of a market game is never null;
(2) while the core is usually not unique, increasing the number of agents in the game tends to eliminate some allocations from the core, so that the size of the core is smaller for larger games, as in the example;
(3) the supply-and-demand equilibrium allocation and ratio of exchange is always a member of the core.

This is a striking result. It says that, for games of exchange, the noncooperative game defined by competitive markets yields a result in the core, and that a great multilateral contract by way of the grand coalition could not improve on bilateral trade mediated through competitive markets. In the language of implementation theory, markets implement the core for this class of games.

8.2.2 Telser on the Core in Games with Production

When we introduce production into the game, the case is quite different. If there are increasing returns to scale, it is quite likely that the core will be null. On the other hand, decreasing returns to scale seems prima facie to conflict with superadditivity, and decreasing returns in the neighborhood of the efficient imputation can also result in a null core. Even if the core is not null, the competitive equilibrium may not be an element of the core, and the core may require a "natural monopoly," with price discrimination or some other measure to efficiently recover overhead costs (Telser, 1978, pp. 129–31). Following Telser we will again assume transferable utility and denote candidate solutions as imputations rather than allocations.

From a mathematical point of view, it is reasonable to consider a null core as a failure for the theory of the core. A solution concept that can never be satisfied for an important class of problems seems mathematically impotent. For Telser, however, the nullity of the core is an explanatory principle. He writes (1978, p. 65):

These results can improve our understanding of the restrictions that are necessary for an equilibrium . . .. These constraints assume a variety of shapes in the real world. The state may intervene either by outright ownership of the plants or by regulation of the activities of the single firm supplying the outputs from its plants. Sometimes the state intervenes by acting on behalf of the buyers, or the buyers may form their own coalition to act in concert . . . in the case of a natural monopoly or a natural monopsony.

A probable example of a null core is the airline industry. Telser (1997 pp. 5–7) gives an illustrative example with two airline companies and just three consumers. However, the conditions of this "toy" model are recognizable in the airline companies of the real world: a very high proportion of costs are overhead costs and profits depend on filling a high proportion of the seats. Telser demonstrates that, in his small-scale example, the core is null so long as competition among the airline companies is unrestricted. Another example along similar lines is the moving picture industry. Telser describes (1997, p. 263) the arrangements by which producers controlled the showing of their films by theaters in the period 1920–1940. These arrangements were considered collusion in restraint of trade and were abandoned under a federal consent decree in 1940. Telser argues (1997 p. 264) that they were optimal, however, as a means of stabilizing what would otherwise have been a game with an empty core.

Despite his passing comment that "Sometimes the state intervenes . . .," Telser is primarily interested in restrictions on competition that arise within the private sector, as the moving picture industry examples show. Nevertheless his ideas supply a key resource for the understanding of public regulation. In general, public regulation may improve the functioning of the economy when it operates to prevent a class of transactions that, if permitted, would result in an empty core. If we adopt the stability interpretation of core theory, we would say that in such a case regulation operates to stabilize an economy that would otherwise have no stable state. Of course, this is not the only function of public regulation, which may be necessary (when there are externalities) to avoid economic states that are stable but inefficient. It is, though, a function of public regulation that is less well understood, and the theory of cooperative games in coalition function form is a key tool to understand it.

Telser writes (1997, p. viii) "People facing empty cores try to devise suitable restrictions and rules in order to obtain efficient outcomes." In the absence of public regulation, they may not succeed. The airline example seems instructive. Telser suggests that vertical integration could resolve the empty core in this case: if each airline were owned by a customer, the core would not be null (Telser, 1997, pp. 10–11). In a real world of airlines with more than three customers, this would mean that the airlines come to be operated by consumers' cooperatives. Of course, this has not occurred in the real world, and does not seem likely to.

Consumers' cooperatives can be successful in operating natural monopolies, as the many examples of rural electric utility and telephone service cooperatives in the United States shows. These cooperatives were established not where service was unstable but where it was not profitable enough for investor-owned companies to offer the service. It may be, though, that the investor-owned companies stayed out of the rural markets because they anticipated empty-core instability. Otherwise it is a bit difficult to explain why services that could be operated at a profit by consumer cooperatives would be refused by profit-maximizing investor-owned companies. In any case, government initiative was central in the establishment of these cooperatives.

Taking the stability interpretation of the theory of the core, an empirical prediction would be that the airline industry would have no stable configuration. That seems to agree with the facts. Bankruptcies and reorganizations of the industry seem to continue in the case of airlines over several decades. If free competition has resulted in instability of the airline industry, as seems to be the case, it does not follow that regulation before 1977 was very successful either. Perhaps the regulation (or other public initiative) appropriate to the airline industry has to be considered an unsolved problem.

Telser's discussion suggests that natural monopoly and monopsony, and other empty-core cost conditions, are quite widespread in a modern economy. In such cases regulation in some sense is unavoidable; and the choice (as indeed Galbraith, 1973, argued) is between public and private regulation. Moreover, efficient private regulation in cases of natural monopoly and monopsony will usually involve price discrimination. Conversely, deregulation has often meant, not a competitive market, but increased and unregulated price discrimination. Statistical studies that show lower average prices following deregulation ignore this, and may not be representative of the prices available to smaller or spot-market traders. All in all, we are unlikely to understand regulation or deregulation without the insights of cooperative game theory.

8.2.3   Values, Power and Accounting

Important applications of the Shapley value include the measurement of power in committees and governments (Shapley and Shubik, 1954) and the allocation of shared costs (Shubik, 1962). Shubik's example for cost allocation supposes that two plants share a joint overhead cost. Here is an example with some similar features to illustrate Shubik's argument.

Table 8.2   Game 8.4: coalitions and values in a game among colleges

Coalitions	Values
∅	0
{E}	−5
{A}	−15
{B}	−5
{A,B}	5
{E,B}	15
{E,A}	5
{E,A,B}	25

West Philadelphia University is a coalition of an engineering school, E, a school of media arts, A, and a business school, B. Each requires the support of a school of arts and sciences, which (for simplicity) generates no tuition revenue. In addition there are other overhead costs such as a computer center. Any coalition, including a singleton (stand-alone school of engineering, art or business) must bear 25 million of overhead costs for these purposes. Operating costs are E, 20; A, 25; B, 5. Tuition revenues are E, 40; A, 35; B, 25. The coalition function is shown as Table 8.2. Suppose, for example, we take the order E, A, B in assembling West Philadelphia University from its parts. The engineering college then must (as the first and so stand-alone unit) bear the overhead alone and generates a value of −5, which is 5 less than no university at all, so $v\{E\} − v\{\varnothing\}$ is −5. Adding an art school creates {E,A}, so $v\{E,A\} − v\{E\} = 5$. Now, add B, creating {EAB}, worth 25. Therefore $v\{E,A,B\} − v\{E,A\} = 20$. Proceeding in this way, considering all possible orders and the appropriate weights, we find that the Shapley values for this game are 10, 1.67, 11.67.

The Shapley values in this case are net of all costs – essentially the target profitabilities of the three colleges, after the overhead cost. The allocations of the shared cost will be the allocations that leave each college with its target profitability. Table 8.3 shows the revenues, operating costs, shared cost allocations and profitability targets (Shapley values) for the three colleges. As we see, in this case the overhead cost is allocated equally, even though the operating costs, tuition revenues, and profitability targets are quite different.

There have been a few other applications of Shapley values to cost assignment, with examples such as the divisions of the Tennessee Valley Authority and the different classes of aircraft that use an airport. This has not been adopted as general accounting practice, but it does provide (at least in principle) an objective standard for the sharing of common fixed costs.

Table 8.3 Allocation of costs, revenues, and profitabilities for three colleges

	E	A	B
Revenue	40.00	35.00	25.00
Operating cost	20.00	25.00	5.00
Allocation	8.33	8.33	8.33
Net	11.67	1.67	11.67

8.3 CHAPTER SUMMARY

Much of the literature of cooperative game theory relies strongly on simplifying assumptions that originate with von Neumann and Morgenstern. There are a number of properties one might like a solution to have: needless to say, no one solution will have all of them. One of the most common solution concepts is the core, which rests on the idea that no group can be denied the payoffs they could obtain if they formed a coalition and chose a joint strategy. The core may, however, be null or may have many potential solutions within it. Three solution concepts that are never null and are unique are the Nash bargaining solution, the Shapley value, and the nucleolus, though the bargaining solution is inapplicable to more than two agents. Applications of these models are largely in economics (and to some extent in political science) and include a theory of exchange, a theory of restrictions on competition, measurement of power and the allocation of shared costs.

NOTES

1. This appears to be inconsistent with the previous sentence, as Telser (1978) notes in a similar context. We might say that the valuations are subjectively consistent in that each coalition is equally (and utterly) pessimistic about the decisions of those outside the coalition.
2. Some of the literature would use the term *preimputation* at this point, and consider it as an *imputation* only if every agent obtains at least what he would get as a singleton. However, that distinction will not be made here.
3. This may occur because the techniques of production involve division of labor (Smith, 1776/1994; Kaldor, 1934), so require a certain minimum workforce to be put into effect.
4. This is demonstrated in the game theory literature by forming a set of axioms one of which is the property, and showing that these axioms are equivalent to the solution concept.
5. These are technical terms from mathematical analysis and will not be discussed in detail here.
6. This follows Forgo et al. (1999).

7. This listing follows Peleg and Sudhölter (2003).
8. This listing follows Peleg and Sudhölter (2003).
9. Extension to divisible goods, drawing in the economic concept of a preference system, would demand a bit of mathematics.
10. For this discussion, an *allocation* of available goods and services among individuals replaces an *imputation* of value to a coalition. The term "allocation" may replace "imputation" in some other applications, following the example of games of exchange.
11. In some contributions this latter assumption is relaxed.

9. Recall, rationality and political economy

It has been observed that much literature in game theory relies on simplifying assumptions that can frustrate the application of the theory, particularly to public policy. The objective of this chapter is to give arguments why several other assumptions are problematic and to sketch some possible alternatives. We will begin with a common (often tacit) assumption of noncooperative game theory and then proceed to explore two further issues of cooperative game theory and an ambiguity in the concept of rationality.

9.1 "BEHAVIOR STRATEGIES SUFFICE"

We now have the technical apparatus to reconsider the role of contingent and behavior strategies in game theory, and the idea that, thanks to Kuhn's demonstration, "behavior strategies suffice." As we recall from Chapter 3, Kuhn had demonstrated that an important family of games in extensive form can be analyzed by using behavior strategies only, choosing local (generally randomized) best responses. It was noted, however, that this analysis is not applicable to games of imperfect recall (CGT, 1997, pp. 146–68), nor to any cooperative game (Selten, 1964), nor does it recover all Nash equilibria (Selten, 1975). It was also stated in Chapter 3 that Kuhn's reasoning does not apply to noncooperative equilibrium concepts other than the Nash equilibrium. This will now be discussed.

In particular, correlated strategies cannot be derived from the local determination of behavior strategies as best responses at each information set. Consider Game 9.1, shown in extensive form by Figure 9.1. No "story" or application will be given for this game, which is offered strictly to illustrate the relation between contingent and behavior strategies. The agents are a and b and the game proceeds in just two stages. First, a chooses between behavior strategies u, c, and d, and then (depending on a's play at the first stage) b chooses between t1 and b1 (at information set B1) or t2 and b2 (at information set B2).

As usual, agent a's contingent strategies need not be distinguished from

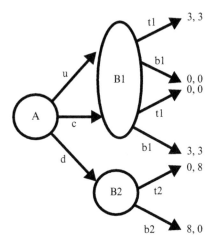

Figure 9.1   Game 9.1 in extensive form

his behavior strategies, since he makes the first play. Agent *b* has four contingent strategies:

1. If *u* or *c* then *t*1, else *t*2
2. If *u* or *c* then *b*1, else *t*2
3. If *u* or *c* then *t*1, else *b*2
4. If *u* or *c* then *b*1, else *b*2

   Notice that this list is highly redundant, as always when behavior strategies are translated to contingent strategies. The reason for this redundancy is that contingent strategies 1 and 3 differ only with respect to play at information set *b*2, which is never reached in Nash-equilibrial play, and similarly strategies 2 and 4. Notice also that any cooperative solution to this game will correspond to a strategy of d by player 1, followed by any behavior strategy of agent *b* and an offsetting side payment. But this can never be realized if behavior strategies are chosen as local best responses.

   Game 9.1 in strategic normal form is given by Table 9.1. This game has a number of Nash equilibria, due to the redundancy that has been mentioned, but they fall into two categories: pure strategy equilibria yielding 3,3 and randomized strategies yielding expected values of 1.5, 1.5. The pure strategy equilibria provide relatively good outcomes but, as usual, they raise questions since they seem to require consultation or information that the agents are assumed (in Figure 9.1) not to have. However,

Table 9.1   Game 9.1 in strategic normal form

Payoff order: A, B		B			
		1	2	3	4
A	u	3,3	0,0	3,3	0,0
	c	0,0	3,3	0,0	3,3
	d	0,8	0,8	8,0	8,0

this game has a simple correlated equilibrium solution. Let the two agents flip a coin and play according to the rules:

For agent *a*: "If *H* then *u* else *c*."

For agent *b*: "If *H* then 1 else 2," or "if *H* and (*u* or *c*) then *t*1 else *t*2 else if *T* and (*u* or *c*) then *b*1 else *t*2."

What a correlated strategy does, in effect, is to imbed the game in a larger game in which the first step is the signal, in this case flipping the coin. Neither agent has any reason to deviate from play by these rules: in *a*'s case, a deviation to d or to play the "wrong" strategy from u and c will leave him with nothing, and in *b*'s case, a deviation to play *b*1 on *H* or *t*1 on *T*, a "wrong" behavior strategy, will similarly leave him with nothing. However, whether the original game is expressed in contingent or behavior strategies, the strategies of play in the larger game must themselves be contingent strategies. To be specific, unless agent *b* knows that agent *a* will play according to the contingent strategy "if *H* then *u* else *c*," agent *b* has no information that would allow him to choose a behavior strategy that would produce a pure strategy equilibrium. This is the logical issue in coordination games, and local choice of behavior strategy offers no escape from it. Local choice of best-response behavior strategies cannot produce a correlated equilibrium in this example.

9.2   EXCHANGE GAMES AND IMPERFECT RECALL

The study of exchange games is one of the great successes of cooperative game theory. Nevertheless, monopoly remains mysterious in the context of exchange games (Aumann, 1973). The conventional monopoly result from the economics textbook, which is inefficient, cannot be consistent with any cooperative solution. Stuart (2001) resolves this problem by imposing the "law of one price," but this assumption is ad hoc, and it is not clear that the law of one price is consistent with cooperative game theory.

The literature of exchange games, like almost all traditional cooperative game theory, assumes perfect recall, although the assumption is never explicitly stated. In this literature, the value of a coalition corresponds to the potential gains from trade among the members of a coalition. Since exchange games are usually modeled as NTU games, even this coalitional value cannot be expressed as a scalar; but in any case no explicit account is likely to be given of the choice of strategies in an underlying game. We will find that if recall is imperfect, the value of a coalition may be less than the potential gains from trade among its members. This section will explore the point by means of a small-scale example.

To assess the implications of imperfect recall for exchange games it will be necessary to be more explicit about strategies in an exchange game. Presumably the behavior strategies in an exchange game are transfers of particular quantities of goods or of money. Certainly each person's agreement to an exchange is an agreement to make a *conditional* transfer and hence is, in a broad sense, a contingent strategy. In order to model exchange as a game in extensive form, we might think of it as a two-stage game. At the first stage each party to the exchange makes a revocable commitment to transfer certain amounts of goods and/or money to certain other parties on the condition of certain transfers from them. At the second stage, if the conditions are consistent then they are carried out, but if the consistency condition has not been met, the offer is revoked and no exchange takes place. The contingent strategy then is of the form "Commit to a transfer of X to i and if i does not commit to a transfer of at least Y to me, then revoke; otherwise do not revoke." This will be called the game of transfers.

With that background consider the following four-person game, Game 9.2. The four agents are of four different types and are denoted a, b, c, and m. Player m is initially endowed with four widgets and his payoff depends on the number of widgets and the quantity of money he has at the end of the game. The payoff is

$$y_m = \begin{cases} 225 \text{ if no widgets are sold} \\ \text{money} + 50z \text{ where } z \text{ widgets are not sold} \end{cases}$$

Agents a, b, and c are initially endowed each with 100 units of money and their payoffs are

$$y_a = \begin{cases} \text{money} + 100 \text{ if one widget} \\ \text{money if no widgets} \end{cases}$$

$$y_b = \begin{cases} \text{money} + 70 \text{ if one widget} \\ \text{money if no widgets} \end{cases}$$

$$y_c = \begin{cases} \text{money} + 40 \text{ if one widget} \\ \text{money if no widgets} \end{cases}$$

It will be apparent that agent *a* will be better off to acquire a widget at a price of less than 100 and *b* at a price of less than 70; however *c* is not benefited by a widget at any price above 40, while *m*'s reservation price of a widget is 50. Note also that if agent *m* were to sell four widgets at the reservation price of 50 each his payoff would be 25 less than it is if there are no sales at all. This 25 is an overhead cost of entering into the market.

Now suppose *m*, *a*, and *b* were to form a coalition and commit themselves to the following strategies:

1.m   *m*: Transfer 1 widget to *a* provided *a* transfers 60 monetary units to me AND Transfer 1 widget to *b* provided *b* transfers 60 monetary units to me.

1.a   *a* and *b*: Transfer 60 monetary units to *m* provided he transfers 1 widget to me.

On this basis the payoffs would be 220 for *m*, 140 for *a*, 110 for *b*, and 100 for *c*, for a total of 570, an efficient outcome. However, with a payoff of 220, *m* is worse off than he would have been with no trade, and so will refuse the coalition.

Suppose instead that the strategies were

2.m   *m*: Transfer 1 widget to *a* provided *a* transfers 82.5 monetary units to me AND Transfer 1 widget to *b* provided *b* transfers 60 monetary units to me.

2.a   *a*: Transfer 82.5 monetary units to *m* provided he transfers 1 widget to me.

2.b   *b*: Transfer 60 monetary units to *m* provided he transfers 1 widget to me.

This yields the same efficient 570 of final payoffs but *m*, *a*, and *b* are each better off with the trade than without, so the coalition might be formed on this basis. From the point of view of the economist, *m*, a monopolist with a fixed cost, has increased his total revenue (in order to cover the fixed cost) by means of price discrimination.

However, strategies 2.a and 2.b are dominated by

3.a   *a*: Transfer 80 monetary units to *b* provided he transfers 1 widget to me.

3.b   *b*: Transfer 60 monetary units to *m* provided he transfers 1 widget to me AND Transfer 1 widget to *a* provided *a* transfers 80 monetary units to me.

Thus, *m* sells only 1 and the payoffs are 210 for *m*, 120 for *a*, 120 for *b*, and 100 for *c*, for a total of 550. This is an inefficient outcome, but *a* and *b* are both better off than they were in the previous case.

The monopolist faces competition from his own customers when he practices price discrimination in this example. He might forestall this by making his strategy

4.m   *m*: Transfer 1 widget to *b* if and only if BOTH [*b* transfers 60 monetary units to me AND *b* makes no transfers of the widget to other players] AND Transfer 1 widget to *a* provided *a* transfers 82.5 monetary units to me.

An alternative in this small game would be to make both transfers of widgets contingent on the money transfers from both players, but in a larger game, in which the monopolist might not be aware of the types (different demands) of the various players, this might not be feasible. On the other hand we see many real-world examples of monopolists attempting to prevent the resale of their products, such as sale contracts that make air tickets non-transferable, anti-scalping laws, and frequent revision of textbooks.

If the game of transfers is a perfect-recall game, then commitment 4.m, with 2.a and 2.b, will result in the same efficient and mutually beneficial outcome as 2.m, 2.a and 2.b as noted above. Suppose, however, that the game is of imperfect recall so that *m* cannot verify that *a* and *b* are committed to 2.a and 2.b rather than 3.a and 3.b. Then strategies 4.m are not available to a coalition of exchange. In such a case we may observe

5.m   *m*: Transfer 1 widget to *a* provided *a* transfers 87.5 monetary units to me.
5.a   *a*: Transfer 87.5 monetary units to *m* provided he transfers 1 widget to me.

In this case *m* makes no offer to *b* because any price acceptable to *b* would, via secondary transfer from *b* to *a*, make it impossible for *m* to obtain an overall revenue that would cover his overhead as well as marginal cost. This results in final payoffs of 237.5 for *m*, 112.5 for *a*, 100 for *b*, and *c*, for a total of 550. This is again an inefficient outcome, but *a* and *m* both find it preferable to no trade, and the others lose nothing by it.

This is a case of monopoly restriction of output such as may be found in any principles of economics textbook. The monopolist has restricted sales in order to charge a high price, and the inefficient payoff of 500 rather than 520 reflects a "deadweight loss" or "welfare triangle" equivalent to

20 monetary units. This is a consequence of "imperfect recall" within the cooperative coalition for exchange.

Suppose that we double the game, so that there are two identical agents of each kind, and again assume perfect recall. The value of a coalition of all active agents $\{m_1, m_2, a_1, a_2, b_1, b_2\}$ will be 965. This is more than double the value of $\{m, a, b\}$ in the previous game because of economies of scale: one of the two producers can supply all four of the active buyers, and the other remains inactive, contributing a value of 225 to the coalition. As before, agents of type c are dummies, contributing their singleton value of 100 to whatever coalitions they join. Now consider a payoff schedule with $y_{m_2} > 225$ and suppose a coalition $\{m_1, a_1, a_2, b_1, b_2\}$ is formed, that is, m_2 is expelled from the grand coalition. The new coalition will increase its value by $y_{m_2} - 225 > 0$, so that the payoff schedule with y_{m_2} is not in the core. For a core allocation, each producer must obtain exactly its alternative cost, 225. By introducing a second producer, and thus a possibility of what economic theory would call market entry, we have introduced a "contestability" constraint (Baumol et al., 1982). An "established monopolist" (the active producer) faces the risk of entry that could precipitate losses or force him out of the market. To forestall this, the established monopolist must moderate his prices to the point that the potential entrant cannot increase his profits by entry. In this symmetrical case, that means the monopolist's own economic profits are reduced to zero. The result is efficient (because of price discrimination) and the buyers benefit from the economies of scale.

Suppose instead that recall is imperfect in the doubled game. On the one hand, by the same reasoning as before, the payoff to each producer must be exactly its opportunity cost, 225. On the other hand, this game is large enough that a producer can profitably sell to agents of type b despite the law of one price. A price consistent with these two constraints is 56.25. There is no monopoly inefficiency, although this is a result of the simplicity of the model, and with additional buyer types – such as a type d who can benefit by buying only at a price of 55 or less – the usual monopoly inefficiency will be restored, on the assumption of imperfect recall.

The conclusions are that, on the one hand, the "law of one price" and monopoly as it is customarily treated in economics are phenomena of "imperfect recall," so we should not expect any counterpart to them in conventional cooperative game theory in which perfect recall is assumed, and (on the other hand) the results of exchange games may be sensitive to deviations from the assumption of perfect recall. We may modify other assumptions, such as introduction of a second potential producer and consequent contestability constraints, and this will change the bargaining power among the agents in the game; but monopoly inefficiency and the law of one price are to be expected only in case imperfect recall is assumed.

9.3   TOWARD POLITICAL ECONOMY

Political economy is an integrated study of economics and politics that allows us to pose and tentatively to answer both positive questions such as "whose interest is state action likely to advance" and "what are the likely consequences of such-and-such economic policy" and normative questions such as "what policies would best advance the interest of the whole population?" Since the formation of coalitions is a foundation both of politics (states, parties) and economics (business firms) the theory of cooperative games in coalition function form might suggest itself as a language of political economy. A difficulty is that we will have at least two games being played simultaneously: a political game and an economic game. Moreover (as in social mechanism design) the *outcome* of the political game is the *set of rules* that define the economic game. The economic game is imbedded in the political game. Imbedding, however, is appropriate to noncooperative games, in that it presupposes that agents are unable to make commitments in the political game that can be relied on in the economic game.

For this section, we will apply *imperfect recall*. The game is supposed to be played in three stages. At the first stage, political decisions are made, determining the "rules of the economic game." At the second stage, production takes place, and at the third stage, exchange. Each participant plays as three "agents" as Kuhn uses the term, respectively in the three stages.

As we have seen, imperfect recall can play a part in neoclassical monopoly theory; indeed, it seems to underlie the "law of one price," which is fundamental also to the neoclassical theory of perfect competition and to classical value theory. Perfect competition is usually considered a noncooperative theory, but perhaps it could better be regarded as a theory of production and exchange with imperfect recall. In any case, to capture the idea that outputs are sold in perfectly competitive markets, imperfect recall will be assumed: any commitments not to resell, which might support monopolization, will be "forgotten" in the exchange game. This enables us to treat the exchange game as "imbedded" in the production game, even though both stages are cooperative. Conversely, the production game determines the endowments for the exchange game.

Second, agents playing in the political game will have no recall of commitments made in the political game when they play the other two stages of the game. Thus, for example, political bribes cannot occur, since the economic agent will not recall that the political agent has offered them. It seems that when we assume imperfect recall in this context we are assuming a certain kind of perfection in the political process. However, this does not go so far as assuming the veil of ignorance at the social contractarian's original position (Rawls, 1971) since the political agent recalls his interests, forgetting only commitments made at the earlier stage of the game. Again, however, we may treat the later games as imbedded in the political game, even though all are cooperative. Even though the agents do not recall commitments from the earlier games, they can anticipate the rational decisions that will be made in the later games; so that the three-stage game is solved by backward induction, with the exchange game as the basic game.

We have seen that a complex economy can result in a null core. This section will revisit Classical Political Economy, which supplies, among other things, a simple technology: a Ricardian corn[1] economy.

9.3.1 A Corn Economy

In the Ricardian framework agents are of three types: landlords, laborers and (capitalist) farmers. For a landlord or capitalist, to seek employment as a laborer is a dominated strategy, but laborer types have no choice, as seeking employment is the only strategy in their strategy sets.

In this model the exchange game comprises exchanges of corn for labor services (employment contracts) and of corn for land services (rental contracts). For the exchange game, landowners are endowed with specific quantities of land and each laborer with a specific quantity of potential labor services, or, in Marxist terms, labor power. Capitalists are endowed with some quantity of corn retained from past production. This stock of corn constitutes the "wages fund,"[2] which limits total wages, since wages must be paid in advance. Payment of rents, however, may be in the form of promissory notes against corn to be produced. The solution will serve to determine the prices, that is, rents and wages. Classical economics suggests the following conjectures: (1) The grand coalition will support one or more imputations in the core of the exchange game. (2) The law of one price will prevail for wages for imputations in the core. (3) If landholdings are of different productivity, not all landholders will receive positive rents. The landholders excluded will be the holders of the least productive land. Other landholders will receive net payments equal to the differential productivity of the land they hold. (4) The rate of profit on stock will be equalized among the different capitalists.

Since corn will be treated as the medium of exchange, inflation and unemployment will play no part in this illustrative model, as in any classical one-commodity model. Problems of macroeconomics will be entirely beyond the scope of the example.

9.3.1.1   The exchange game

There are m players of landlord type, w players of worker type, and c players of capitalist type, indexed with landlords $1, \ldots, m$; capitalists $m + 1, \ldots, m + c$; workers $m + c + 1, \ldots, m + c + a$.

In a classical corn economy, wages are limited by the wages fund, which is the stock of corn in the hands of capitalists. For this example, each capitalist is endowed with a stock of corn amounting to b units, and plans to employ u workers. For the purpose of the example, for simplicity, capitalists are homogenous. The wages fund is cb, and consequently the average wage is $cb/w = b/u$. Each landlord is endowed with a plot of land suited for cultivation by u workers. (This defines a unit of land.) The productivity of these plots varies, and the output of plot i, if it is cultivated, is $q_i u$. (If a single person owns more than one plot of land then he can enter the game as a separate player for each plot.) The plots of land are indexed in such a way that $i < j \Rightarrow q_i \geq q_j$. It is assumed that $m > w/u$. Each worker is endowed with a certain amount of potential labor and nothing else. Then a coalition of one capitalist with u workers and one landlord is (assuming the average wage is paid) the smallest that can engage in production and has no redundant resources. There are c capitalists with $cu = w$.

Definition. Now consider a coalition of w' workers, c' capitalists, and m' landlords comprising a set $L = \{i, j, \ldots, k\}$. Let $r = \min (m', w'/u, c')$ and let $M \subset L \ni |M| = r$ and $i \in M, j \in L \backslash M \Rightarrow i < j \Rightarrow q_i > q_j$. That is, if the quantity of land is more than can be used with the labor that the coalition includes and can pay, then only the most productive plots will be used. If $r < m'$, the coalition is called a redundant-land coalition. In any case, the value of the coalition is $v = \Sigma_{i \in M} q_i$.

From this definition it follows that the exchange game is superadditive. This being so, we need consider only the grand coalition, since any stable imputation will be admissible for the grand coalition. Schmeidler's (1969) terminology will be adapted as follows. Let C be a deviation from the grand coalition, that is, a coalition that proposes to act separately, and y_i an imputation for the grand coalition. Then $V(C) - \Sigma_{i \in C} y_i$ is the *excess* for C. It follows that y_i is in the core if the excess for any deviation C is nonpositive.

Since by assumption $m > w/u$ the grand coalition is a redundant-land coalition. In a Ricardian model the "marginal land" is the most productive land not currently in use. In this model, then, plot $(w/u) + 1$ is the marginal land and plots with $i > w/u + 1$ are inframarginal. Here are some inferences about this exchange game.

(1) Marginal and inframarginal land receives no rent. Consider landlord i with $i > w/u$. Suppose the payoff to landlord i, $y_i > 0$ and consider the deviation to $C = N\backslash\{i\}$. Then $v(C) = v(N)$, so the excess for this deviation is exactly y_i; and it follows that the grand coalition with a positive rent for a parcel of marginal or inframarginal land is not an element of the core of the game.

(2) For any set of one capitalist i and u workers $\{j, \ldots, k\} = L$, $y^* = y_i + \sum_{j \in L} y_j = q_{(w/u+1)}$. That is, a capitalist and a standard team of workers together receive exactly the productivity of the marginal land. Let $S = \{i, j \ldots k, w + 1)$.

    a.   Suppose instead that $y^* > q_{(w/u)+1}$ Consider the deviation to $C = N\backslash S$. The excess for this deviation is $y^* - q_{(w/u)+1} > 0$, so a candidate solution with $y^* > q_{(w/u)+1}$ is outside the core.

    b.   Suppose instead that $y^* < q_{(w/u)+1}$. Consider the deviation to S. The excess for this deviation is $q_{(w/u)+1} - y^* > 0$, so a candidate solution with $y^* < q_{(w/u)+1}$ is outside the core.

(3) The law of one price applies to wages. By the above result, if any two workers j, k have different payoffs $y_j \neq y_k$, the substitution of one for the other in the elements of S would result in different values for the sum of the payoffs to S, contrary to proposition 2.

(4) The rate of profit is equalized. For any two capitalists i, j we must have $y_i = y_j$ by the same reasoning. Moreover, the common rate of profit is $\pi = (q_{w+1} - b)/b$.

(5) Rent is differential productivity. Consider a landlord with $i \leq w$. Then the payment to this landlord is $y_i = q_i - q_{w+1}$. Let S be $\{i, j, \ldots k, l\}$ where $j, \ldots, k$ are any u workers and l any capitalist. Suppose $y_i < q_i - q_{w+1}$ and consider a deviation to $N\backslash S$. The value of the deviation is $v(N) - q_i$. The payoff to $N\backslash S$ in the candidate solution is $V(N) - (y_i - q_{w+1})$ (by 2 above) so that the excess is $y_i - (q_i - q_{w+1}) > 0$. Suppose $y_i < q_i - q_{w+1}$ and consider the deviation to S. The excess for this deviation is $q_i - y_i - q_{w+1} > 0$. Thus a payment to a landlord that differs from the differential productivity will not be an element of the core.

(6) The core comprises the grand coalition or other coalition structure with payments of differential rent to landlords, ca/w to workers, and $q_{w+1} - a$ to capitalists. Consider any deviation S with w' workers, c' capitalists, and m' landlords comprising a set $L = \{i, j, \ldots k)$. The payoff to this group in the grand coalition is $y = (cb/w)w' + (q_{w+1} - b)c' + \sum_{i \in L}(q_i - q_{w+1})$. The value of the deviation is $v(S) = \sum_{i \in M} q_i$ with M as in the definition; and with r as in the definition this can be rewritten as $(cb/w)r + (q_{w+1} - b)r + \sum_{i \in L'}(q_i - q_{w+1})$. Term by term, each of these terms is less than or equal to the corresponding term in

y. Thus the excess is nonpositive and the deviation does not disrupt the candidate solution.

Thus, the exchange game in this case will generate predictable prices, rents, and profits. Agents in the production game will anticipate this, and make their decisions accordingly. We now proceed to consider the production game.

9.3.1.2 The production game

Players in the production game are agents with enough wealth to supply their own (negligible) consumption of corn and to command the labor of a team of workers. We may assume that relations between employees and workers in production are noncooperative and that employers face a cost of monitoring effort, and that this cost includes both an overhead component and a variable component that increases with the size of the workforce. We suppose that a workforce of just u workers is the team size that optimally balances the overhead cost against the variable cost. Thus, each production coalition will plan for a workforce of just that size.

In a corn economy, the formation of a coalition for production can be identified with the formation of capital through frugality, or obtaining capital by alienating land. Classical economics assumes that landlords never save nor employ labor in cultivation. Nevertheless they command purchasing power in the form of corn, from their rental income. In a classical model, they spend their income on the wages of servants. These servants are consequently not available as employees in the exchange game as just described, and constitute the "unproductive labor" in the classical schema.

Accordingly, we suppose that agents in the production game are of two types, with different intertemporal preferences: one type with a lower discount rate for future consumption and one with a higher discount rate. An agent of the first type, endowed with land, and an agent of the second kind, endowed with capital, might form a coalition to transfer the capital to the agent of the first type in exchange for land at a price in corn between the agents' discounted present values of future rents. We also suppose that any attempt to consolidate production on two or more plots of land results in loss of efficiency due to "imperfect recall," so that production coalitions with more than one capitalist will be unstable. Classical economics did not envision externalities, so they will play no part in this model. This is a counterfactual simplifying assumption. As noted, a property-owner with more than enough land or labor to put one work team to work in cultivation will enter the exchange game as a plurality of agents, one for each plot to be cultivated, and this is the stable partition for the production game.

9.3.2   The Government Game

Now consider the government game. We will take as given that the government is formed as a grand coalition of all players in the government game. Marx (who after all originated the concept of capitalism as an economic system) regarded capitalist government as a "dictatorship of the bourgeoisie," that is, a system in which laborers would be excluded from participation in politics by a property test for the franchise or some other unprivileged status. (This does not conflict in any way with the views of Smith, Malthus, and Ricardo.) Thus, for Marx (1848), "The executive of the modern state is but a committee for managing the common affairs of the whole bourgeoisie." Accordingly, laborers are not players in the government game. The governing grand coalition of capitalists and landlords is not an agent in the production or exchange games. While further formal development will be beyond the scope of this sketch, it seems that a government of proprietors would serve their interests by establishing a regime of property rights and largely unrestricted contracts, while mercantilist measures, which in practice favor some proprietors but not others, would be a subject of controversy.

9.4   RATIONALITY

The contrast has been noted again and again between contingent strategies underlying cooperative game theory and the behavior strategy often applied in noncooperative game theory; and so also has the contrast between cooperative and noncooperative game theory in general. In this section the argument will be made that these differences are at base different conceptions of rationality. The issue is *not* whether people are rational, irrational or partly irrational. Rather, the issue is what it means to say that people are rational, either wholly or partly.

9.4.1   Weakness of Will and Rationality

Noncooperative game theory has been greatly influenced by Selten, and his 1975 paper is a key paper for our purposes here. It is interesting to contrast the assumptions of this paper with Selten's earlier view (1964). There, we recall, Selten had acknowledged in an afterword that his (cooperative) model in the 1964 paper required the assumption that players could commit themselves to any contingent pure strategy, acknowledging Thomas Schelling (1960) as the source of his change of mind. In 1975 Selten's assumptions are reverse to those he made in 1964, but his terminology is also inconsistent in a way that may obscure the difference. Selten defines a pure strategy not as a plan assigning probability 1 to one of all possible contingency plans, as von Neumann and Morgenstern did and as Selten did in 1964, but as the assignment of probability 1 to a particular behavior strategy choice at a particular information set. Selten now assumes what Schelling (1980) called weakness of will.

But this will make no difference for rational behavior as Selten now conceives it. Selten limits his subject matter to games with perfect recall. He writes (1975 p. 320): "Since game theory is concerned with the behavior of absolutely rational decision makers whose capabilities of reasoning and memorizing are unlimited, a game, where the players are individuals rather than teams, must have perfect recall." He then excludes consideration of teams, and he justifies this by limiting his scope to "strictly noncooperative games." This means Kuhn's multiple-agent games are excluded. For multiple agents to be joined together as a single player would require a cooperative agreement among them, and this is excluded by assumption. But this, taken with Selten (1975 p. 328): "Player 2's choices should not be guided by his payoff expectations in the whole game but by his conditional payoff expectations," tells us that for rational behavior as Selten now conceives it, there can be no commitment whatever. By assumption, only behavior strategies are relevant to rational behavior.

This concept of rationality has become predominant in economics as well as noncooperative game theory, and it is appropriate now to expand on the different concepts of rationality in those fields and in most cooperative game theory.

In economics, the issue of weakness of will and commitment arises in the context of intertemporal inconsistency of rational choice (Strotz, 1956). We adopt the neoclassical convention of expressing time preference by a discount rate. Most economic literature assumes that this discount rate per unit time is the same regardless of the delay before the payment is made. This assumption of a uniform rate of time preference has no basis in empirical observation, but is made in order to reconcile the theory of rational choice, as it is understood in modern economics, with the assumption of time preference. The difficulty is that a non-constant rate of discount can result in what are called intertemporal inconsistencies in decision making. What this means is that a rational, maximizing decision maker would make one decision at one point of time, but at a later point of time would rationally prefer the alternative he has initially, rationally rejected. (There has been some recent research on alternatives to constant rates of time preference, such as hyperbolic discounting, but it has been directed to a different issue.)

9.4.2　Intertemporal Inconsistency

Let us illustrate this point by an example. Suppose that the decision maker discounts any prospect delayed by more than six months at 18 percent, but that his rate of discount for prospects delayed six months or less is zero. Now the decision maker must choose at t_0 between two alternatives. Alternative A1 is a payment of $5000 at $t_0 + 1$ year. Alternative A2 is a payment of $10000 at $t_0 + 5$ years, but A2 has a cancellation clause: at any time during the first year, for a cancellation fee of $100, the decision maker can cancel his decision for A2 and receive the payment of $5000 at $t_0 + 1$ year.

At t_0, the discounted present values are:

Alternative A1　$4237
Alternative A2　$4371

Accordingly, the decision maker chooses alternative A2. However, at $t_1 = t_0 + 6$ months and one day, the payoff for alternative A1 is less than six months away, and so is not discounted, and is valued at $5000. To obtain this payment, however, the decision maker must pay the cancellation fee of $100. The net values discounted to t_1 are

Alternative A1　$4900
Alternative A2　$4748

Therefore, the rational decision maker reverses his decision.

This is a one-person game. Suppose we express these decisions as plans of action for the successive stages like the pure strategies as understood by von Neumann and Morgenstern. The decision maker has three pure strategies:

(1)　Choose A1.
(2)　Choose A2, then do not cancel.
(3)　Choose A2, then cancel.

The payoffs of these strategies, discounted to t_0, are:

(1)　$4237
(2)　$4371
(3)　$4127

Why, then, does our rational decision maker not simply choose strategy 2 and stick with it? Suppose that the decision maker has a weak will, in

Schelling's sense, and knows that he does. Then he can anticipate that if he chooses A2, he will indeed cancel it after six months and in fact carry out strategy 3. Because of his weakness of will, strategy 2 simply is not available to him. That being so, in the spirit of Ulysses and the Sirens, (note Elster, 1977) the rational but weak-willed decision maker will choose strategy 1 and alternative A1.

This is not to say that intertemporal inconsistency does not exist. No doubt a strong-willed decision maker, having chosen strategy 2, will feel some subjective tension in the nature of regret or temptation during the time interval t_1 to $t_2 = t_0 +$ one year. Does rationality require him to act on the temptation? Well – perhaps it does.

9.4.3   Weakness of Will in a Game in Extensive Form

Weakness of will may also be a factor in interactive decisions. Consider Game 9.3 in extensive form, shown as Figure 9.2. All decisions are close enough together in time that there is no need to discount payments to present value.

First we note that the perfect equilibrium for this game is for decision maker *a* to choose behavior strategy Alt1 for a payoff of $4237. However, when we express this game in terms of contingent strategies, we have, for decision maker *a*,

(1′)   Choose Alt1.
(2′)   Choose Alt2, then, if *b* chooses up, choose up.
(3′)   Choose Alt2, then, if *b* chooses up, choose down.

and for decision maker *b*,

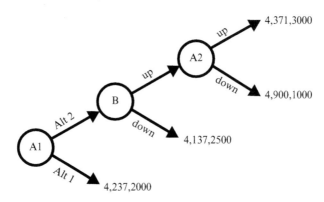

Figure 9.2   Game 9.3: two-person game

(4')   If a chooses Alt2 then choose up
(5')   If a chooses Alt2 then choose down

If decision maker a chooses strategy 3', then decision maker b's best response is strategy 5', while if decision maker a chooses strategy 2', *and b knows this with certainty*, then b's best response is strategy 4'. Taking this into account, the payoffs to be expected from these strategies would seem to be

(1')   \$4237
(2')   \$4371
(3')   \$4137
(4')   \$3000
(5')   \$2000

This being so, we ask again, why does decision maker a not simply choose strategy 2'? There are two possibilities: (1) Decision maker b believes that decision maker a has a weak will, and will not carry out strategy 2 but, having arrived at decision-point $a2$, will choose down. Decision maker b therefore chooses strategy 5'; and this is known to agent a, who then chooses strategy 1' as his best response to strategy 5'. Thus, it seems, the subgame perfect equilibrium can be necessary because of the belief that a has a weak will. (2) The second possibility is that b believes a is dishonest and opportunistic and will choose "down" at decision point A2 regardless of any protestations to the contrary. Thus, the subgame perfect equilibrium can be necessary because of the belief that A is dishonest. But suppose that decision maker A has a strong will, that is, a capability to choose strategy 2' and stick to it despite the temptation to choose "down" at decision point A2; and suppose that this is known to decision maker b. Suppose decision maker a also is honest, and this, too, is known to decision maker b. Thus, decision maker needs only announce "on my honor, I am choosing strategy 2'," and then b's rational decision is for strategy 4', and the cooperative solution A1, up, up results.

9.4.4   Perfect and Ideal Rationality

Notice that, so far as a's decisions are concerned, decision sequence 1', 2', 3', with its payoffs, is identical to 1, 2, 3 in the intertemporal inconsistency example. The major difference is that it is decision maker b's belief that a has a weak will, rather than a's belief that a's will is weak, that puts Alt2, up, up out of a's reach. Supposing b to be rational, what basis might he have for that belief? One possibility is that we *define* rationality as maximization *constrained by weakness of will*. Then we need only apply common knowledge of rationality to induce *b*'s belief in *a*'s weakness of will. I submit that this is indeed the concept of rationality noncooperative game theory and in neoclassical economics. In what follows choices that maximize payoffs subject to the constraint of weakness of will be called *perfectly rational* choices, not because their outcomes are perfect (as the example shows) but because it is rationality in this sense that defines subgame perfect equilibrium.

But common knowledge of *perfect* rationality is not the only possibility, and we need to consider others. First consider the possibility that *b* believes *a* is dishonest. Then *b* will not believe any assertions by *a* that he will choose "up" at decision point A2 and accordingly *b* chooses strategy 5′. But (1) *a*'s honesty is of concern to *b* only if *b* believes *a* has strong will. If *b* believes *a* has a weak will then *b*'s decision will not be affected by the further knowledge that *a* is honest or dishonest. (2) *a* can benefit by acting dishonestly only if *b* believes both that *a* has strong will and that *a* is honest. (3) Accordingly, we must consider a 4-step game in which *a*'s decision whether to act honestly or dishonestly is the first stage. If *a* has a strong will he can commit himself to one or the other and carry out the commitment. (4) However, if *b* believes *a* has chosen to act honestly, then *a*'s best response is dishonesty. (5) Therefore, this first stage requires a mixed-strategy solution. (6) Since *b* is rational, he will be aware of this and will accordingly estimate the payoffs of strategies 4′ and 5′ as expected values reflecting the optimal mixed strategy for *a*, which is to act honestly with probability 2/5. Thus, *b*'s belief that *a* will be dishonest with probability 1 is either irrelevant or irrational.

Common sense suggests that rationality, strength of will and honesty are distinct traits and that rational individuals may exist in positive numbers whose will is strong and is weak; and that within each category some are honest and some are crooked. These conditions are also relative and a typical person is more likely to act in an honest and strong-willed way in some circumstances than others. Suppose *b* believes that a very large proportion of all human beings have weak wills, but has no way to know which type *a* is. In that case, once again, he would estimate the payoffs of his choices as expected values, using the probabilities based on the frequency of weakness of will in the population and such other evidence as he may have. To fail to do so would be irrational or at best boundedly rational! Perfect rationality is naïve on this score, and in what follows decisions based on maximization with estimates of the probability that other agents have strong wills and are honest will be called "sophisticated rationality" to distinguish them from the rationality expressed by subgame perfect equilibrium. (For the concept of sophisticated rationality and some evidence that there are multiple types of decision makers in a real human population, see Stahl and Wilson, 1995.)

It seems that b's behavior, as assumed in subgame perfect equilibrium theory, can be rational only if b believes that weakness of will is a common trait of all human beings. This in turn can be considered a rational belief only if (1) it is true, or (2) b's experience has been so idiosyncratic that it seems to b that the belief is true, although b is mistaken. We can eliminate (2) as inappropriate to be the basis of a general theory, and conclude that: for subgame perfect equilibrium theory, universal weakness of will is a necessary assumption. If both weakness of will and perfect rationality are common human characteristics, then there is little point in distinguishing between them. But the results of such an identification can be rather peculiar. The results of the example of intertemporal inconsistency and of the two-person game from Figure 9.2 can both be stated in the following way: (1) Define rationality as perfect rationality. (2) Suppose decision maker a in fact adopts strategy 2 (or 2') and carries it out. (3) As a result of this choice, decision maker a is better off. (4) Decision maker a has acted irrationally. Stated in just that way, perfect rationality is not a very intuitively appealing concept of rationality.

How would von Neumann and Morgenstern have treated the game in Figure 9.2? In the first instance they would have expected the two players to form a coalition around strategies 2', 4', since the total value generated by that pair, $7371, dominates all other strategy pairs. This will be no difficulty if both have strong wills and are honest.

In Game 9.3, the noncooperative equilibrium is also the assurance value for both players. Unlike Game 6.8, for example, Game 9.3 has no threat strategies. For a game like Game 6.8, von Neumann and Morgenstern seem to envision a negotiating process along the following lines: agent a says "if you adopt strategy D2 I will adopt strategy P, leaving you with 5 rather than 7." This is a threat designed to increase b's bargaining power, and for von Neumann and Morgenstern (and Nash to the contrary) *all feasible threats are credible*. But none of this makes sense unless each agent believes the other has strength of will enough to carry out his threats, even when they are irrational in the sense of perfect rationality. For von Neumann and Morgenstern, a rational agent maximizes his expected utility on the assumption that all agents maximize and all have strong wills. Strength of will is here considered an aspect of rationality. In what follows, rationality in this sense will be called "ideal rationality."

Notice that the assumption that all feasible threats are credible is central to the definition of the coalition function in von Neumann and Morgenstern, and as most cooperative game theory is based on the characteristic function, we may conclude that the assumption of ideal rationality is characteristic of cooperative game theory. Thus, it is appropriate to distinguish between cooperative and noncooperative game theory by noting that while noncooperative game theory assumes perfect rationality, cooperative game theory assumes ideal rationality.

9.4.5 Bounded Rationality

We now have three concepts of rationality: perfect, ideal and sophisticated. Ideal rationality is not perfect, perfect rationality is not ideal, and neither is sophisticated (if some agents in the actual world have strong wills and others do not). Moreover, there is reason to believe that none is very descriptive of actual human behavior. Computation of a perfect, ideal or sophisticated rational solution to a problem may require a great deal of cognitive effort, and cognitive effort is a very scarce resource for real human beings. Acting according to a rule of thumb that may not be "optimal" from any point of view is "boundedly rational." Can "boundedly" rational decisions be ideal, perfect or sophisticated?

The answer is yes. Suppose that in fact the population comprises individuals both with strong and weak wills, and this is a known fact. Then only sophisticated rationality can be defended as consistent with rational belief. Suppose then that an agent faces a threat, and will attempt to judge the credibility of the threat. Suppose also that the individual believes, on evidence and experience, that most people (though not all) have weak wills. Then computing a perfect equilibrium for the game will be easier than computing a sophisticated solution, since the sophisticated solution requires us to know the perfect solution anyway (and in a particular case, such as Figure 9.2, the perfect solution may be very easy indeed). Then the rule that "only subgame perfect threats are credible" is boundedly rational. Suppose in addition (as seems very plausible) that most people are better able to exercise a strong will in some circumstances than in others, and that in particular, the agents are situated in a culture that values personal honor highly and regards oath-breaking as dishonorable. In such a society, we suppose, the probability that an oath will be carried out is very high. Then the ideally rational solution may be boundedly rational, in case oaths have been sworn. Finally, consider the game of "running the red light." A stoplight is a correlated equilibrium solution to an anticoordination game: its perfectly rational solution is to obey the light. However, if an individual is ideally rational, he may commit himself to running the stoplight, and carry the strategy out despite the temptation to stop at the last second. This could maximize his utility if the light has just changed and his intention is made very clear by speeding up. Having observed that about one in three drivers will do this (in Philadelphia) the sophisticated solution of delaying one's start into a green light (without trying to see whether the driver coming the other way has speeded up or not) is boundedly rational, although it is neither ideal nor perfect.

9.4.6 Perfect Rationality and the Manipulation of Elections

Let us return to Gibbard's theorem on manipulation of voting (Chapter 7, section 7.3) As we have observed, it belongs to noncooperative game theory. Indeed Gibbard remarks (p. 593): "If a system does make outcomes a function of preferences, it is in virtue of individual integrity, ignorance, or stupidity." An interesting example arises from the US presidential election of year 2000. In that election, about 3 percent of the popular vote was cast for the Green Party candidate, Ralph Nader, and it is quite likely that if those votes had gone to Gore, George Bush would never have become president. Some Democrats criticized the Green Party voters very bitterly, with the implication that they acted dishonorably by failing to cast a strategic vote for Gore. Now, one could argue (consistently with Gibbard's model) that the "honest" vote in an American presidential election is a vote for whichever of the "realistic" candidates one prefers. But this example may be better understood in the light of cooperative game theory.

In elections prior to 2000, Democrats and Greens had constituted an informal de facto coalition. (American politics does not allow formal coalitions, with the partial exception of New York State.) In the period before the 2000 election (in effect) the Greens demanded a bigger payoff from the coalition, threatening that if their demands were not met, they would vote for a third party candidate, taking whatever risk of a Republican (or worse) victory as might follow. The Democrats, perhaps assuming that the Greens would act with perfect rationality – or that however strong their will might be the Greens were bluffing – declined the demand. The Greens would "execute the threat[,] ... not ... something [the Greens] would want to do, just for itself" (Nash, 1953, p. 130). In so doing the Greens acted with ideal, but not perfect, rationality. The Democrats' failure was a failure of sophisticated rationality: they failed to take adequate account of the mixture of perfectly and ideally (and sophisticatedly) rational agents in their interactive decision problem.

It seems clear that Gibbard's model presupposes perfect rationality. In a world of ideally rational (and honest) agents, it would be quite possible for a grand coalition to arrive at a "social contract" to vote honestly, and come to a general agreement as to what that might mean. "Individual integrity" would then assure that collective decisions would depend only on honestly expressed individual preferences. (Perhaps we should not assume too hastily that this would be a good thing.) In the absence of such a social contract, though, it is not clear that nonmanipulated voting has any meaning at all in a cooperative approach to game theory. Is not every political coalition an attempt to manipulate the outcome? In a real world in which there are at least a few ideally rational agents, some of whom are honest sometimes, the only real rationality is sophisticated rationality. But this hardly favors a case for nonmanipulated voting! What does seem clear, nevertheless, is that models based on perfect (or ideal) rationality alone are likely to mislead us. The very concept of a coalition, after all, derives from (European) electoral politics, and to try to discuss political choice in noncooperative terms that exclude most coalitions seems odd, if very American.

9.4.7   Coalition Formation

Suppose that the population includes individuals both with strong and weak wills, and at least some of those with strong wills are honest. How then will coalitions form? First, there will be no mutually beneficial coalitions comprising only the weak-willed. Such coalitions would accomplish nothing that would not be accomplished by a noncooperative equilibrium. The typical coalition, then, will include at least a subset of strong-willed individuals who adopt threat strategies that encourage the others to keep their agreements and correlate their strategies so as to increase the value of the coalition. These strong-willed individuals may be known to the others as leaders, but more probably as officious busybodies, nosy parkers, or snitches. It may be that the officious busybodies, nosy parkers and snitches will form a grand coalition and formalize some of their threat strategies as institutions such as property rights and enforcement of contracts. If, as seems likely, people are better able to act with strength of will in some circumstances than in others, we are likely to see cooperative arrangements more often in some social circumstances than others, for example among people of a common religious faith, and to see a good deal of noncooperative interaction among the coalitions that do occur. If a part of the population are both strong-willed and dishonest, they may be able to form some coalitions for their dishonest purposes, by means of committed threat strategies, even though dishonesty breeds distrust and distrust is an obstacle to cooperation, as in Game 9.3. For coalitions of this kind, the "irrationality" of gang vendettas would be seen as an expression of ideal (though not perfect) rationality. We recall criminal gang leaders with nicknames like "Buggsy" or "Banannas," nicknames that express their lack of (perfect) rationality.

It does seem likely that both cooperative and noncooperative game theory are mistaken in their extreme views. On the one hand, commitment does occur in human interactions, and on the other hand, it is not easy nor altogether predictable. It seems more plausible to say that real human beings can make and carry out commitments – that is, some people can, sometimes, and under some circumstances! We may then suggest some circumstances that favor successful commitment:

(1)   The existence of a contract enforced by a third party such as the state or a private bondholder.
(2)   Repeated interaction with other parties, over a long term.
(3)   Patience.
(4)   Strong relevant social norms of promise keeping or honor.
(5)   A high trust environment.
(6)   An agreement consistent with motives of equity or reciprocity.
(7)   A large differential between the payoffs attainable by cooperative action and those that result from noncooperation.

No doubt other such circumstances can be offered.

9.5   SUMMARY

This chapter has reconsidered several aspects of noncooperative game theory, beginning with subgame perfect equilibrium and the analysis of games in extensive form. First, the analysis of such games in terms of local optimization of behavior strategies returns only a subset of Nash equilibria and thus may not return correlated equilibria. Correlated equilibria presuppose contingent strategies in some cases.

Second, analysis of games in extensive form following Kuhn's paper in Kuhn and Tucker (1953) and Selten (1975) assume perfect recall. However, for some purposes, imperfect recall may be a more productive assumption. Imperfect recall allows us to recover the neoclassical model of monopoly in a cooperative analysis, to treat the economic game as being (ideally) imbedded in a political game, and provides a characterization of "perfect" competition that is an alternative to the assumption of a continuum of traders.

Finally, there is a contrast of two concepts of rationality. "Ideal rationality" links rationality to strength of will. It seems that ideal rationality is characteristic of cooperative game theory and is the substantive difference that distinguishes cooperative game theory from noncooperative game theory. The rationality of Selten (1976) is perfect rationality. "Perfect rationality" links rationality to weakness of will. The example of intertemporal inconsistency shows that perfect rationality characterizes neoclassical economics as well as noncooperative game theory. A third concept of rationality is "sophisticated rationality," which is consistent with the belief that the population includes both types with strong and with weak will. This belief leads toward a world very much like the world we seem to live in, a world not susceptible to analysis in terms either of perfect or of ideal rationality.

NOTES

1. Recall that "corn economy" is a phrase in the British language, not the American, so that "corn" means "grain," such as wheat or oats, depending on the country, rather than maize specifically. For this section "corn" is an abstract wage-good.
2. This wages fund assumption was perhaps one of the least persuasive of classical ideas, and the first to be abandoned, but contains an important true insight: if production takes time, it will be necessary for the people who do the work to eat while production takes place, and if the money wage is raised while there is no increase in the quantity of wage-goods available, inflation is the only result. This fact has reasserted itself in the context of industrialization in the twentieth century and as recently as the food crisis of the spring of 2008. The unique function of capitalists in the classical corn economy is to supply the wage-goods from their stock of accumulated wage-goods.

PART II

Mixed cooperative and noncooperative decisions: extensions

10. Biform games and considerable solutions

The previous chapter has argued that cooperative and noncooperative game theory reflect different conceptions of rationality – ideal and perfect, that is, cooperative and noncooperative rationality respectively. Neoclassical economics and game theory within the scope of the "Nash Program" (Chapter 3, section 3.2 above) assume that human rationality is fundamentally perfect, and deviations from that (including what might appear to be ideal rationality) are to be explained by the particular circumstances, such as enforceable contracts with given penalties for violation. We might explore the opposite hypothesis: that human rationality is fundamentally ideal, and deviations from ideal rationality are to be explained by the particular circumstances. As Nobel Laureate Eric Maskin has observed (2004) "We live our lives in coalitions." Among these "coalitions" are business firms, political organizations, churches, clubs, unions, and perhaps governments. In conventional economics and noncooperative game theory, these economic and other organizations are analyzed in terms of noncooperative decisions, contracts and organizational rules, a genre exemplified, for example, by the work of Jean Tirole, honored by the 2014 Nobel Prize and that of Oliver Williamson, honored by the 2009 Nobel (Royal Swedish Academy of Sciences). Even if economic organizations are fundamentally cooperative, this work has much to teach us about the ways human beings whose rationality is neither ideal nor perfect may attempt to approximate a cooperative arrangement. Nevertheless, it seems worthwhile to model economic and political organizations as fundamentally cooperative, that is, as arrangements among ideally rational decision makers. After all, if the purpose of contracts and organizational rules is to approximate a cooperative arrangement, it could be useful to know what a cooperative arrangement might be.

10.1 A FRAMEWORK FOR A THEORY OF THE MIDDLE GROUND

There are, however, some major obstacles in the way of an applicable theory of this kind. Ideal rationality is explored in cooperative game theory. Much of the literature of cooperative game theory relies on a set of simplifying assumptions that are rarely applicable to the actual world of economics and politics. The predominant tradition in cooperative game theory is the theory of superadditive games in coalition function form, that is, games in which:

(1) Each person is a member of exactly one coalition.
(2) The value created by the coalition depends only on the members of the coalition, and not on any influence from outside the coalition, nor on the distribution of the benefits of the coalition among the members.
(3) There can be no advantage in decentralization. As a consequence the grand coalition of all decision makers will form, and thus:
     a. No noncooperative decisions will be observed;
     b. Inefficiencies will not be observed, even in cases such as externalities;[1]
     c. A range of threat behaviors, such as holdouts from contributions to public goods, are excluded by assumption.

Each enumerated point is contradicted by our experience of economic organizations. Further, if we nevertheless adopt them all, we cannot be sure that there is a satisfactory solution. The core may be a null set – that is, there may be no set of coalitions and arrangements that is stable under the pressure of competition. If we relax (2), and allow the value created to depend on the distribution of benefits among the members of the coalition, we enter the realm of nontransferable utility, NTU games. With these assumptions we can obtain a satisfactory theory of exchange, as noted in Chapters 3 and 8. However, when production is introduced, empty cores typically result. And when the core is not null, it will often have a continuum of imputations. This raises the core assignment problem: given infinitely many outcomes that are stable under competition, which one does the theory specifically predict? For this purpose a distinct theory of core value imputation or bargaining is needed, and bargaining theory remains in something of an unsatisfactory state.

Thus, the theory of superadditive games in coalition function form will not serve the purposes of this part of the book. What seems to be needed is a theory of the middle ground, which admits both cooperative and noncooperative decisions in specific cases, and in which some coalitions, including the grand coalition, may be impossible to form. Some useful hints can be found in the literature on search and matching in labor markets. While this literature treats decisions as mainly[2] noncooperative, it provides a theory of the *formation* of mutually beneficial (employment) relations through a costly process of information-gathering. Suppose, then, that decisions are made with ideal rationality, but that cooperative decisions require more and richer information than noncooperative decisions, and this information is costly. Then, for some decisions, the cost of information necessary for cooperative decisions will be greater than the benefit of cooperative decisions, so that noncooperative decisions are consistent with ideal rationality. This approach has been explored at length, but for the most part informally, in McCain (2014a).

Even if cooperative coalition values are superadditive, the coalitions that can be formed will be limited by the information landscape, so that the grand coalition and some other coalitions cannot form. Since the decision to share information must itself be made noncooperatively, we may adopt the formalism of biform games, in which the first stage of the game is noncooperative while the payoffs are determined by a second, cooperative stage. While null cores cannot be excluded in principle, the informational cost of forming new coalitions will increase the probability of core stability by limiting the formation of dominating coalitions. This will not be formally modeled in this book: instead, we will simply focus on stable (core) allocations where they exist. But, conversely, the cost of coalition formation will by the same token increase the probability that the stable solutions comprise a continuum (as indeed they do in search-and-matching labor market models, for example Mortensen and Pissarides, 1994). Thus a bargaining theory will be required. This book will draw on two models from McCain (2013): for bargaining, Chapter 6, and the Feasible Coalition Structure (FCS) game, Chapters 8–9.

Following that outline, this part will envision a world in which:

(1) Coalitions take place on a population that is dispersed but among whom there are linkages. The linkages take the mathematical form of a tree or forest, as the terms are used in graph theory.[3] The coalitions that can be formed are limited to groups who are linked. In principle, more or less indirect linkages might be sufficient, but for present purposes it will be convenient to limit that somewhat and to allow coalitions only within a group all of the members of which are linked to a single member (see McCain 2013, p. 192).

(2) The links are links of information-sharing for cooperative decision making, and may be of more than one kind, depending on the information that is shared.

(3) In the absence of links, the information necessary for cooperation is not available, and interactions can only be noncooperative.

(4) Candidate solutions are considerable solutions. In a considerable solution, coalitional decisions are Pareto-optimal with respect to the members of the coalition and Nash-equilibrial as among distinct coalitions.

    a. On the one hand, informational links that would facilitate cooperative arrangements will in general not exist between the distinct coalitions; that indeed is the reason they remain distinct coalitions.

    b. When an individual is a member of two or more coalitions, she or he will support changes in the strategies of one of the two coalitions that increase her or his payoff only to the extent that they do not reduce her or his payoff in the other coalition. Thus, rationality conditions for individuals who are members of two or more coalitions point toward the Nash equilibrium solution for decisions of the two coalitions with shared membership.

    c. For a more formal discussion see McCain (2013, pp. 178–9, 187).

(5) Decisions take place in a four-step sequence.

    a. The first step is the formation of links, at some resource cost. This is unavoidably at least partly noncooperative and costly. At this first step, also, some links may spontaneously disappear.

        i. This adapts ideas from the search-and-matching literature in macroeconomics.

        ii. For links between a coalition and a potential employee, there will be a component of the productivity of the employee in the coalition that is particular to the match of the employee to the coalition, that is, an *idiosyncratic productivity shock*. The formation of the link makes the value of the idiosyncratic productivity shock known to all involved parties.

        iii. For links with potential customers or capital partners, idiosyncratic shocks may affect other aspects of the coalition's strategy.

    b. The second step is the formation of coalitions within the links that remain from the first stage or from earlier periods.

    c. The third step is the determination of a common strategy for the members of the coalition.

    d. The fourth step is the distribution of the value created according to bargaining power.

(6) This sequence can be solved by backward induction, which in this case is backward induction from the cooperative stages to the noncooperative stage, consistently with Brandenburger and Stuart's (2007) "Biform Games." If we may treat the coalition in terms of transferable utility, then the induction back from the fourth stage to the previous stages is straightforward. For a more general non-transferable utility treatment, the third and fourth stages may not be clearly distinguished and may need to be treated simultaneously.

10.2   BIFORM GAMES

For this model, some decisions are made cooperatively and some noncooperatively. There is some work in the literature that draws on both streams of literature in a coherent way. Greenberg's *Theory of Social Situations* (1990) and Zhao's hybrid solutions (1992) provide instances. Brandenburger and Stuart (1996, 2007) have proposed a model that draws on well-known ideas from both streams of game theory in a coherent and relatively simple way. Their model is denoted as "biform games" and will be used in this and subsequent chapters.

A biform game is a two-stage game. The first stage is noncooperative and is usually represented in normal form. In a noncooperative game in normal form, each set comprising the strategies chosen by the players in the game results in a schedule of payoffs to the players. For a biform game, each set comprising the strategies chosen by the players in the game instead results in a cooperative game to be played by the players.[4] The cooperative game to be played is one of several, and the selection depends on the list of strategies chosen by the players in the noncooperative stage. In the language of Chapter 2, we may say that the *outcome* of the noncooperative play is one of a list of cooperative games. The biform game as a whole then is solved by backward induction, with a cooperative solution at the second stage that then determines the payoffs to the noncooperative first-stage game.

The biform game formalism rests on an idea mentioned by von Neumann and Morgenstern in passing: that the decision to enter into a coalition with other specific agents can itself only be made noncooperatively (1944/2004, pp. 222–3). Biform games were conceived in the context of the economics of corporate strategy, and the applications discussed by Brandenburg and Stuart (1996, 2007) and others (Stuart, 2001; Chatain and Zemsky, 2007) have been in that context. In that context, of course, competition is a key determinant, and accordingly the cooperative game solution concept used is the core. In other contexts, perhaps, other cooperative game solution concepts might be more appropriate; but that will have to be a subject of future study, and this book also will apply the core or closely related concepts at the second stage.

When the core is non-unique, bargaining theory or some similar algorithm may be needed to identify specific payoffs.[5] With this qualification, biform games can be used to model some familiar problems, including monopoly (Stuart, 2001), externalities, costly job matching, and corporate governance (McCain, 2013), and issues of business strategy such as branding for intermediate products (Brandenburg and Stuart, 2007) and the choice of the scope of a firm's offering (Chatain and Zemsky, 2007).

Table 10.1   Game I: values with the advanced technology

v{*a,b*}	20
v{*a*}	12
v{*b*}	8

Here is an example to illustrate the approach. Two firms, *a* and *b*, are considering producing complementary products, such as software and hardware for some computational system. Each can commit in advance either to an advanced technology for the product or instead to a proven, but less advanced technology. These are the strategies in their noncooperative game. For Firm *b* the commitment to the advanced technology will be more expensive than it will be for Firm *a*. If both commit to the advanced technology, this technology can also be used in the finished computational system, and that will expand the market for it. In any case, if they do not cooperate on the product, then they may find other markets for their expertise in the advanced technology. Thus, if both commit to the advanced technology, the second-stage cooperative game will be Game I as shown in Table 10.1.

If, by contrast, *either* of them fails to commit to the advanced technology, then it is not available for the final product, nor is there a market for the more advanced expertise in case the two firms do not form a coalition. Thus the game is Game II, shown in Table 10.2. In both cases the core of the second-stage cooperative game is unique, so that the first-stage noncooperative game can be expressed as in Table 10.3. This game has a

Table 10.2   Game II: values with the established technology

v{*a,b*}	10
v{*a*}	5
v{*b*}	5

Table 10.3   The first-stage noncooperative game

First payoff to Firm A, Second to Firm B		Firm B	
		commit	don't
Firm A	commit	12,8	5,5
	don't	5,5	5,5

dominant strategy equilibrium where both firms commit and the payoffs are split as 12,8. Of course, this example has made some simplifying assumptions, especially to assure that the core value assignment would be unique.

For a second example, consider again the noncooperative game shown at Table 4.1. Two neighboring towns choose strategies to meet their needs for water, but they have different opportunities. On the one hand, Downstream can build two dams and Upstream only one; but on the other hand, Upstream's dam will interfere somewhat with any dam Downstream may build. We suppose that they must commit themselves to building dams first before forming any coalition for cooperation in water supplies. For some reason, they cannot communicate until the dams are built, but thereafter, they can form a cooperative water coalition if there is benefit from doing so. Thus, their strategies lead them to play one of games I–VI in Table 10.4.

Games I–IV and VI will be rather simple. In I–III, Upstream has enough water for their own use, but none to share. Thus there will be no mutual benefit from a cooperative arrangement: the value of the coalition of Upstream and Downstream will be just the sum of the values of the singleton coalitions of the two towns acting separately, that is, 6, 5, and 6 respectively. For IV and VI, Downstream has no spare water to share, although Upstream would benefit from it, so again there is no mutual benefit from a coalition and the values of the two-town coalitions are respectively 6 and 2. However, Game V is different. Downstream has water to share and Upstream is needful of more water. In Section 4.1.3, the payoffs of 4,4 in this case were justified by the idea that the dams built by Downstream would supply water also to Upstream, a cooperative arrangement. Thus the value of the two-town coalition is 8. But what if they do not agree? Then Upstream has insufficient water, for a payoff of 1, while Downstream bears the cost of two dams, leaving a net payoff of 3. Thus, Game V is as shown in Table 10.5.

For Games I–IV and VI, the core will be unique, and corresponds to the

Table 10.4 Game 4.2 reconsidered: Water Works

Payoff order: Downstream, Upstream		Upstream	
		one dam	no dam
Downstream	one dam	I	IV
	two dams	II	V
	no dam	III	VI

Table 10.5   Game V: sharing costs and water

v{Upstream, Downstream}	8
v{Upstream}	1
v{Downstream}	3

values that each of the towns can create independently, that is, the payoffs in Table 4.2. However, the core of Game V is not unique. Any schedule of payoffs that gives at least 1 to Upstream and 3 to Downstream is in the core of the game, since neither town can do better by dropping out of the cooperative arrangement. There is a surplus of 4, 50 percent of the total, that can be distributed between the two towns in any proportion that may result from their bargaining power. Suppose Upstream has three times the bargaining power of Downstream, so that the surplus is distributed as 3 to Upstream and 1 to Downstream and then we have the payoffs shown in Table 4.2, 4,4. (Perhaps the Mayor of Upstream is the boss of a political machine that can retaliate by refusing to appoint residents of Downstream to public service jobs.) Suppose instead that the two towns have equal bargaining power, so that the surplus is divided equally, 2 to each. Then the payoffs would instead be 5, 3.

We see that the reduced first-stage game may depend on the relative bargaining power of the cooperating agents, or on some other influence that determines the division of any surplus. (Brandenburger and Stuart 2007 propose a model based on optimism and pessimism, which are supposed to be given characteristics of the agents.) In this case, with the assumption that Upstream has ¾ of the bargaining power, we see the payoffs at the noncooperative stage that were considered in Section 4.1.3. For this game, we recall, the Nash equilibrium is the inefficient one at which each town builds one dam. Thus there will be no mutually beneficial coalition. How does this inefficient outcome occur? Recall, we assumed at the outset that the two towns could not communicate until after the dams were built. Artificial as this assumption may be, it points to a valid possibility: inability to communicate in a timely fashion may result in inefficient equilibria. Another possibility is lack of trust: Upstream might expect that even if they had an agreement, Downstream would opportunistically choose their dominant strategy of building only one dam to enjoy a payoff of 5. But either way, part of the problem derives from Upstream's preponderant bargaining power. If the two towns had equal bargaining power, recall, the payoffs for Game V would be 5,3, and this would be one of two Nash equilibria in the noncooperative game. A rational agent might indeed refuse to enter into a cooperative group because she or he could anticipate that the bargaining power of others would lead to a payoff less than she or he could obtain independently. In any case, the result of a biform game analysis (that relies on the core as the solution concept at the second stage) will often depend on bargaining power as in this instance.

The biform game formalism thus allows for both noncooperative and cooperative decisions, and that is what we observe in the actual world. In particular, inefficient phenomena such as externalities and unemployed resources may enter as a consequence of the preliminary stage of noncooperative decisions. (For examples see McCain, 2013, Chapter 7, section 7.3.5).

10.3   BARGAINING AND THREAT

Since bargaining solutions will be particularly important for this part of the book, some comments on the history and status of bargaining theory will be needed. The first modern bargaining theory is found in Zeuthen's *Problems of Monopoly and Economic Warfare* (1930), in his discussion of wage bargaining. Letting y_1 and y_2 be the agreed payoffs to bargainers 1 and 2, and supposing these payoffs are limited by a possibility frontier $f(y_1, y_2) \leq 0$, and letting v_1 and v_2 be the outside options of bargainers 1 and 2 respectively, Zeuthen says that the bargain will be such as to maximize the product $(y_1 - v_1)(y_2 - v_2)$ subject to $f(y_1, y_2) \leq 0$. For Zeuthen, the only alternative to an agreement is that one or another bargainer withdraws from the negotiations completely, and he supports the product solution with reasoning about the motivation of each bargainer to avoid such a withdrawal. Thus, the outcome in the case of a bargaining impasse is that the bargainers receive their outside options, and this assumption has dominated bargaining theory in the Zeuthen–Nash tradition since.

Two decades later Nash (1950) obtained the same formula as his solution to the bargaining game, but by a quite different procedure. Nash characterized the solution by a set of axioms that seemed appropriate for a reasonable bargain, and derived the product solution from the axioms. This coincidence of Nash's result and that of Zeuthen, by quite different approaches, must have lent something to the popularity of the solution, although the coincidence was not known until Harsanyi pointed it out in 1956. However, Nash's 1953 paper contains an important step forward that seems to have been somewhat neglected. What is most often remembered about this paper is that he constructs a noncooperative model of bargaining, the Nash Demand Game. However, he observes that this game has a continuum of Nash equilibria. To obtain uniqueness, Nash constructs the first refinement of Nash equilibria, and that refinement yields, as the unique result, the product formula common to his 1950 paper and Zeuthen's book. But Nash offers no plausible argument to support the refinement, which seems suspiciously ad-hoc. Nevertheless, the paper is often thought of as creating some connection between the Nash bargaining theory and the noncooperative Nash equilibrium. More important for present purposes, however, is that in this paper, Nash constructs a variable-threat bargaining model. This is a two-stage model. At the first stage, the bargainers noncooperatively choose their "threats." The threats determine the impasse outcomes of the game: letting s_1 be bargainer 1's threat and s_2 be bargainer 2's threat, then $v_1 = g_1(s_1,s_2)$, $v_2 = g_2(s_1,s_2)$, where g_1 and g_2 are functions known to the bargainers. This two-stage game is solved by backward induction, with the product formula used to solve for the payoffs given v_1 and v_2 for each set of threats, and the threats then chosen noncooperatively, as a Nash equilibrium. Nash then provides an axiomatic derivation of this two-stage model, incorporating the axiomata in his 1950 paper.

Now consider the two sets of threat strategies, $\Sigma_1 = \{s_1^1, s_1^2, \ldots, s_1^q\}$ and $\Sigma_2 = \{s_2^1, s_2^2, \ldots, s_2^r\}$. Presumably, the threat to withdraw from the coalition and go it alone will be a member of each of these two sets of threat strategies, and should either bargainer choose that threat, then v_1 and v_2 will be the outside options. However, the possibility of other impasse outcomes v_1 and v_2, distinct from the outside options, must imply that there can be other threats that do not require the dissolution of the coalition. And in real bargaining circumstances there clearly often are. Strikes – perhaps the most frequent threat strategy in applied bargaining theory – are not ordinarily intended to dissolve the coalition permanently (and the violence with which strikebreakers have often been met is evidence of this). Better examples, perhaps, are found in work-to-rules slowdowns and in the noncooperative withdrawal of effort envisioned in some versions of efficiency wage theory. Even if these threats are carried out, the coalition continues to function, only at some cost to some members. A further issue that arises from these examples is that they take place in coalitions of many more than two members, and are effective only if they are carried out by a group. However, group threats have played little part in formal bargaining theory.

The bargaining theories of Zeuthen and Nash are inconsistent with any differences in bargaining power. Pen (1952), probably unaware of Nash's work, generalized Zeuthen's theory to allow differences in bargaining power by substituting the power function $(y_1 - v_1)^\beta(y_2 - v_2)^{1-\beta}$ for Zeuthen's product solution. In this generalization, "bargaining power" is truly a black box. Later, Roth (1979) and Svejnar (1986) provided axiomatic bases for this generalization, and extended the model to $n > 2$

bargainers and (in the case of Svejnar) provided empirical evidence that a model with differences in bargaining power could fit empirical data better.

A further issue that arises with the examples of strikes and slowdowns is that they take place in coalitions of many more than two members, and are effective only if they are carried out by a group. However, group threats are inconsistent with the two-person bargaining games of Nash, Zeuthen, and Pen and have played little part in formal bargaining theory. But group threats also play no explicit role in the n-person bargaining models of Roth and Svejnar. Perhaps, however, the differing exponents β attached to different bargainers could capture this influence.

Another, more strictly noncooperative, approach to bargaining focused on the bargaining process, and, assuming that bargainers with positive time preference would prefer an agreement to be settled sooner rather than later, derives a solution to the bargaining process on the basis of those considerations. Working outside game theory, Cross originated this sort of model in 1965. Cross was aware of Nash's ideas but did not apply game theory, perhaps because the subgame perfect equilibrium concept was not yet known. Using subgame perfect equilibrium, Rubinstein (1982) derived a game theoretic model from a similar view of bargaining. This gave rise to a large literature, and a widespread impression that Rubinstein had provided a new (noncooperative) justification for the Nash solution. In fact, the Rubinstein solution approximates the Nash solution only in the limit and in a special case, and subgame perfect equilibrium models of the bargaining process give rise to a fairly large range of solutions depending on the special case assumptions one makes. What they all have in common – in the language of threats – is the assumption that the threat that determines the bargaining outcome is the threat of a delay in the bargaining process itself, and in such a case the outside option may have little impact on the bargaining result.

Bargaining models derived new urgency from the emergence of search-and-matching models of labor markets (for example, Mortensen and Pissarides, 1994). In these models, because matching is costly but the costs of matching are sunk, a successful job match creates a surplus. Mortensen and Pissarides adopt the Zeuthen–Nash bargaining model, identifying the impasse outcomes with the bargainers' outside options which, in the case of the employee, depends on the rate of unemployment. But this model fails empirically (Shimer, 2005). The logical root of the problem is that, in these models: (1) the wage is proportional to the employee's outside option, (2) the expected value of the outside option depends on the probability of prompt re-employment, and (3) this probability is assumed to depend in a very sensitive way on the rate of unemployment. Hall and Milgrom (2008) resolve this by adopting a bargaining theory more in the tradition of Cross

(1965) and Rubinstein (1982), that is, one in which the decisive threat is to delay the bargaining process itself. These models all treat the bargain as if there were only two bargainers, one employer and one employee, and thus are not applicable where a larger group can generate a larger surplus nor where group threats may play a part.

Cooperative game theory produced some value solutions that can be interpreted as bargaining models. Of these the best known is the Shapley value, and it has sometimes been interpreted as an n-person bargaining model in which bargaining power is proportional to the "average marginal contribution." For present purposes, the nucleolus (Schmeidler, 1969) is of particular interest, since it allows groups to exercise bargaining power, and is based on an inverse of the net gain relative to an outside option. In this model a group derives bargaining power from the threat to withdraw from the coalition; other threats, either individual or group, play no role. For Schmeidler, only the least advantaged group has bargaining power; but as Forgo et al. (1999) observe, this reflects Schmeidler's aggregation of the net gain from cooperation by a min–max operator. Forgo et al. suggest a minimum sum of squares aggregator to arrive at a more equalitarian, fairer distribution. McCain (2013, Chapter 6) instead aggregates the net gains by means of the power function. This is shown to be a generalization of the Zeuthen–Nash formula. McCain's aggregation allows for differences of bargaining power along the lines of Pen's generalization, with the reasoning that these differences in bargaining power may arise from threats, including group threats such as the work-to-rules slowdown, that do not require the dissolution of the coalition. The derivation of the bargaining power from the threats is not formally modeled, however. This is the bargaining model used in this Part of the book.

10.4   PLAN OF THE PART

This part of the book will outline a model of a market economy that draws together the concepts of biform games, with noncooperative search and matching at the first stage, with considerable solutions and cooperative decisions by an array of coalitions at the second stage, drawing also on the bargaining theory developed in Chapters 6 and 7 of McCain (2013). In the spirit of backward induction, we will begin in Chapter 11 with a discussion of bargaining as a basis for the imputation of coalitional value creation to individual payoffs. That chapter will also discuss the selection of a joint strategy for a coalition that resembles a business firm. Chapter 12 will discuss the formation of coalitions, first from the (cooperative) perspective of the core of the game, and subsequently with respect to the (noncooperative) search-and-matching processes to form new links. Chapter 13 will reconsider the theory of monopoly and monopsony and its policy implications, and will revisit the theory of employment policy from the point of view of the model constructed in Chapters 11 and 12. Chapter 14 reconsiders some issues on the determination of wages. In Chapter 15 the political process itself is considered, again from the point of view of an extended bargaining theory. Legislation is considered as a two-stage process, with bargaining at the first stage and voting at the second stage. This is, of course, a theory of representative government, since it relies in indirect links via elected representatives, among citizens many of whom are not sufficiently densely linked to permit direct cooperative action. Contradictions and sources of failure in this model of the political process are considered. In all of these discussions bargaining power plays a central role.

NOTES

1. Shapley and Shubik (1969) model externalities in the closely related non-transferable utility games, but externalities do not seem to generate inefficiencies in their model.
2. Bargaining theory is introduced, and while the term "noncooperative bargaining" is used, the model applied is Nash's cooperative bargaining model – not his noncooperative "demand game" (for example, Mortensen and Pissarides, 1994).
3. In informal terms, a tree is a set of points with links that may be direct or indirect but do not form any cycles, and a forest is (of course!) a set of discrete trees.
4. For a mathematically more formal definition see McCain (2013, p. 131).
5. On the possibility of an empty core, see McCain (2013, pp. 130–31).

11. The firm as a coalition

For this part of the book, the business firm is envisioned as a coalition in a kind of cooperative game. The members of the coalition are of three types: employees or workers, proprietors, and customers. This chapter has two objectives. First, if we think of the firm as a cooperative coalition then we assume that its decisions are Pareto-optimal as among its members. Thus, one objective is to explore the implications of this assumption, particularly for public policy, and contrast them with the implications of the orthodox theory of the firm. Second, it would be helpful if we could treat the cooperative game of economic organization as a transferable utility game, in which the firm's decisions first maximize the value of the firm and then distribute that value among the members according to bargaining power or some other distributive schema. In such a case we may regard the firm's strategic or allocative decisions and its distributive decisions as independent. Thus, the second objective of the chapter is to identify simplifying assumptions and normalizations that would allow this independence, and to explore the limits and implications of these assumptions.

11.1 PRELIMINARIA

As we envision the business firm as a cooperative coalition among proprietors, employees and customers, a grounding assumption is that cooperative decision making requires intensive sharing of information. Thus the coalition is limited to those who have previously formed links of information sharing. Activities that create information-sharing links are known as job search and recruitment, advertising, marketing, and shopping (Hall, 2008). The decision to engage in these activities can only be made noncooperatively, since *a fortiori* the links necessary for cooperative decisions do not exist when the decision to form such a link is made. Nevertheless, when a person considers whether to seek a new job (for example), she does her best to anticipate the benefits she will obtain from a new affiliation. This is the intuition behind the method of backward induction. Once decisions have been made to seek and make new links, and new affiliations are formed, the decisions to seek new links are past and irreversible, and the costs associated with those decisions are sunk. In order to understand what the job-seeker or marketer anticipates, we must first understand the decisions that will be made at the last, cooperative stage. That is an objective of this chapter.

We also envision a world in which an individual will, as a rule, be a member of two or more coalitions, as employee, customer, or proprietor. Unfortunately, most of the received literature on cooperative games (in coalition function or partition function form) does not address this. Accordingly, this chapter will build on the model sketched in McCain (2013, pp. 184–5). A coalition structure is simply a set of coalitions and is feasible, for present purposes, if appropriate informational links exist among the members of the coalition.

Coalitions are formed with a view to mutual benefit, that is, the potentiality that the coalition, by coordinating the strategies of its members, can realize a surplus over the benefits that they might receive alone or in other affiliations. This surplus will then be distributed among the members of the coalition. The schedule of benefits or payoffs to the members is an *imputation* for the coalition. An imputation is feasible if the total of the payments for each coalition is no more than the surplus generated by that coalition (this follows Aumann and Dreze, 1974). A *candidate solution* is a coalition structure and a set of feasible imputations for the coalitions that make up the coalition structure. A *solution* is a candidate solution that is stable in the face of the efforts of rational agents to increase their benefits from all of the coalitions that they join. In the theory of games in coalition function form, for example, the core is the set of candidate solutions that are stable in that sense.

Not all candidate solutions can be considered as potential solutions by rational agents. The *considerable solutions* are those for which the imputations and strategies of each coalition are (1) Pareto-optimal as among the members of the coalition and (2) Nash-equilibrial among the various coalitions (McCain, 2013, Theorem 9.1, pp. 187–90). The implications of Nash equilibria for public policy are well understood and some of them have been discussed in Part I. This part will have little to add on that point. However, the implications of Pareto-optimality among the members of the coalition may include some novelties. The orthodox "theory of the firm" assumes that interactions among the members of the firm are noncooperative or (at best) some ad-hoc hotchpotch of cooperative and noncooperative elements. Thus the cooperative perspective has the potential of new insights.

To say that a cooperative coalition is formed for mutual benefit is not to say that the benefits will be distributed symmetrically. The imputations to the various members of the business coalition will depend on two things:

the size of the surplus and the relative bargaining power of the members. It may be that these two things are independent, so that we may say that the coalition first maximizes its surplus and then divides the surplus according to bargaining power. Section 11.2 will consider a case in which this simplification is viable, and some assumptions that allow us to implement it. The model in which the maximum value of the coalition is independent of the imputation will be called a *value creation* model. Section 11.3 considers the implications of a value creation model for some policy issues, and finds that they contrast importantly with those derived from orthodox economics. Section 11.4 discusses some limitations of the model.

It has been said that the coalition exists to produce some mutual benefit above what the members could obtain in alternative affiliations. Accordingly the model posits a well-defined second-best alternative or outside option for each member of the coalition. Since the member would not rationally choose to participate in the coalition if she were to receive less, the outside option is often known in cooperative game theory as the rationality constraint for that individual.

The bargaining power value creation model used in this chapter subsumes the orthodox assumption of profit maximization as a special case: for that case we suppose that only the proprietors have positive bargaining power. In the next chapter, however, we will find that profit maximization implies a dynamic inconsistency.

For this model it is assumed that the customer is a consumer, that is, a final user who is motivated by the subjective benefits of use of the coalition's product or service. This excludes business-to-business (B to B) trade, in fact a high proportion of all economic activity. This follows the convention of orthodox economics, which in effect treats the firm as a vertically integrated producer of a consumer good called "value added."[1] This convention is adopted for the purpose of comparison and contrast with the orthodox model. Extension of the value creation model to B to B trade would be simple. The model assumes that nonhuman means of production are supplied by the proprietor. For these purposes, the proprietor of one coalition may also be a member of one or more other coalitions as a customer. Explicit discussion of such a model of B to B trade will be beyond the scope of the book, however.

The treatment of a business firm as a cooperative coalition including proprietors and employees is not new (Aoki, 1980; McCain, 1980) although it has not been much explored. The inclusion of customers appears to be new. (It is suggested by a paper of Hall, 2008, which however assumes that prices are determined noncooperatively.) Now, clearly, customers, employees, and proprietors are agents of different types. The distinction of types of agents is more familiar in noncooperative game theory than in cooperative game theory. Types may be distinguished by the different ways their payoffs are determined for a given set of strategies. Since these differences may be offset by side payments in a cooperative arrangement, they may not seem to be important for a cooperative analysis. However, these distinctions will be important for the investigation in the next section of the chapter, and for the determination of bargaining power.

11.2 A COALITION FOR PRODUCTION AND SALE

In the context of backward solution, the business firm makes production and sales decisions at the last stage, once the membership of the coalition is given. Thus, the firm is a monopolist in respect of its customers and a monopsonist in respect of its employees. Nevertheless, a condition of cooperative decision making is that the decisions are Pareto-optimal as among the members of the coalition, and as we have seen, this condition is incorporated into our model. A first step is to characterize these efficient decisions, the surplus or "value created" by the coalition, and its distribution among the members of the coalition.

A business firm creates value by producing some quantity or array of goods or services for the benefit of the customers. Labor and nonhuman resources supplied by the other members of the coalition provide the means of the production. On these points our model will be very conventional. For simplicity we assume that the firm produces a single output. As in the traditional "short run" analysis, we will take the nonhuman inputs as given and assume that output increases with the labor input and that the marginal productivity of aggregated labor is non-increasing.

Traditionally, economic theory simplifies by identifying the labor input with the head count of the firm's employees. For this discussion we will need a little more detail. A given workforce may increase the labor input by working longer hours or by working faster or more intensely, at the cost of greater effort. We thus suppose that each employee i commits a quantum of effort h_i that is a scalar representation of what might be a complicated vector of hours, speed of motion, attention, and other effortful activities. At the same time we allow for idiosyncratic productivity.[2] The idiosyncratic productivity of employee i is k_i. Thus the aggregate labor resource of the firm is the weighted sum $\sum_i k_i h_i$. (This approach is suggested by a large literature in the microfoundations of macroeconomics; see, for example, Mortenson and Pissarides, 1994.) In equation (A11.1a) in the appendix to this chapter, the labor used in production is limited by $\sum_i k_i h_i - L_0$, where L_0 is a labor overhead comprising that part of the coalition's aggregate labor supply that is allocated to activities other than the production of goods and services to satisfy the wants of customers, including supervisory labor, recruiting and human resource management, and marketing. We assume that effort enters into the utility function of the employee as a source of "disutility" beyond some nonnegative threshold, so that the marginal cost of labor can be identified with the money value of the marginal disutility of labor in a way that is formally quite conventional, though the interpretation differs somewhat from the standard one in the principles of economics. Further, we will assume that each employee is employed by only one firm.

Customers may buy a greater or lesser quantity of the output or service of the firm. As a result they are the direct beneficiaries of the firm's activities and will compensate the other members of the firm with a side payment. We will assume that they have a conventional utility function for the product. We *may not* assume that the law of one price will apply. The familiar Marshall–Lerner conditions (Lerner, 1934) tell us that a price set by a non-discriminating monopolist will be Pareto-inefficient, so these conditions will NOT be applicable in our case. Instead, second-degree price discrimination or some all-or-nothing offer will be the rule.[3] For simplicity, we will assume that the side payment from the customer to the coalition is determined as an efficient price for the quantity consumed together with a lump-sum transfer that may be either positive or negative (from the point of view of the customer). This pricing scheme is commonly called a "two-part tariff" if the transfer is the same for all customers; however, we assume here that they may differ from one customer to another according to bargaining power (see, for example, Goolsbee et al. 2013, p. 429 and note Gifford and Kudrle, 2010, p. 1299). Customers will, in general, be members of several coalitions as customers, and this element of "competition" will affect a customer's rationality constraint.

Following a long tradition in economic theory, we will assume for simplicity that there is just one owner, who receives the residuum after wage payments are deducted from the customers' aggregate side payments. The proprietor provides the nonhuman means of production. Thus all issues of capital structure and the informational cost of seeking new capital partners are avoided, at this stage. Aggregate labor is transformed into output according to a conventional production function $Q = F(L)$.

With an appropriate normalization of the utility functions of the employees and customers, we may identify the gross surplus generated by the coalition with the sum of the consumers' and producers' surplus. This enables us to treat the firm's decisions in terms of a transferable utility game. Thus, we may say that the firm first maximizes its value and, as a second step, distributes the value among its members according to bargaining power.

11.2.1 Normalization

We first consider the utility of employees. Following convention, we suppose that workers are averse to effort at the margin and express that assumption by positing a utility function such as

$$u_i = \Omega_i(\overrightarrow{X}_i, h_i) \tag{11.1}$$

with

$$\frac{\partial \Omega_i}{\partial h_i} < 0, \frac{\partial^2 \Omega_i}{\partial h_i^2} < 0 \text{ for } h_i > h_i^0 \geq 0$$

Here i refers to a particular individual worker, $\overrightarrow{X}_i$ is a vector of quantities of consumption goods, perhaps indexed by date and contingency, and h_i is the worker's effort commitment. As a simplifying assumption, we assume that Ω_i is additively separable (Scitovszky, 1943), that is

$$\Omega_i = \omega_i^*(\overrightarrow{X}_i) + g_i^*(h_i) \tag{11.2}$$

We assume that the coalition's side payment to i is a lump sum payment z_i, based on an all-or-nothing offer. A further simplifying assumption replaces $\omega^*_i(\overrightarrow{X}_i)$ with $\omega_i(z_i)$. For convenience we reverse the sign of $g_i^*(h_i)$ so that (11.2) becomes

$$u_i = \omega_i(z_i) - g_i(h_i) \tag{11.3}$$

Finally, observing that $\frac{\partial \omega_i}{\partial z_i} > 0$ and that utility functions are unique only up to a monotonic transformation, we transform the utility function so that $\omega(z_i) = z_i$, yielding

$$u_i = z_i - g_i(h_i) \tag{11.4}$$

Thus, $g_i(h_i)$ is the "disutility of effort" expressed in money terms. Expression (11.4) is expression (A11.2a) in the appendix to this chapter.

The side payment from a customer to the coalition will often reflect second-degree price discrimination or an all-or-nothing offer or, equivalently, a uniform (marginal) price per unit purchased, pq_ι, together with a lump sum transfer z_ι. Here ι refers to a particular customer, q_ι to the quantity of the coalition's product or service taken by ι, and z_ι to the lump sum transfer from ι to the coalition, which may be either positive or negative depending on the relative bargaining power of ι. Suppose ι consumes differentiated products and services provided by K distinct coalitions 1,

$2, \ldots, K$, where here we are concerned with the payment of ι to coalition 1. The uniform prices set by these coalitions are $p_1, p_2, \ldots, p_K$. We may represent ι's preferences by a conventional utility function or by its dual, the expenditure function

$$E_\iota = E_\iota(u_\iota, p_1, p_2, \ldots, p_K) \tag{11.5}$$

where u_ι is an index of the utility enjoyed by customer ι. In the context of a considerable solution, we treat the prices $p_2, \ldots, p_K$ and lump sum transfers from customer ι to coalitions $2, \ldots, K$ as givens: they are the strategies of other players (coalitions) in a Nash equilibrium among the distinct coalitions. As usual, u_ι can be represented as

$$u_\iota = u_\iota(q_1, q_2, \ldots q_k) \tag{11.6}$$

where q_j is the quantity of the product or service of coalition j that minimizes E given u_ι. Taking $q_2, q_3, \ldots, q_K$ as given in relation to (11.5) and (11.6), let

$$z_\iota^* = E_\iota(u_\iota(q_1, q_2, \ldots, q_K), p_1, p_2, \ldots, p_K) - E_\iota(u_\iota(0, q_2, \ldots, q_K), p_1, p_2, \ldots, p_K) \tag{11.7}$$

Then z^* is the maximum lump sum transfer that ι can make to coalition 1. If required to make a larger lump sum transfer, then ι would be better off to drop out of coalition 1.

Suppose $z_\iota = 0$. Then $E = E^{max}$, corresponding to ι's consumption budget for the current period net of lump sum transfers to other coalitions. For any z_ι such that $0 < z_\iota < z_\iota^*$ the corresponding expenditure is $E_{z_\iota} = E^{max} - z_\iota$ and there is a corresponding utility index u_{z_ι}. For $z_\iota^1 < z_\iota^2, u_{z_\iota^1} > u_{z_\iota^2}$. Thus, by a monotonic transformation of the utility function we may have

$$u_{z_\iota} = z_\iota^* - z_\iota \tag{11.8}$$

For $z_\iota = 0, u_{z_\iota} = z_\iota^*$.

Now we impose the simplifying assumption that the utility function is additively separable as

$$u_\iota(q_1, q_2, \ldots q_k) = f_\iota(q_1) + \Omega_\iota(q_2, \ldots, q_K) \tag{11.9}$$

with $f_\iota(0) = 0$. At $z_\iota = 0$, this is

$$u_0 = f_\iota(q_1) + \Omega_\iota(q_1, \ldots q_K) \tag{11.10}$$

At $z_1 = z_1^*$,

$$u_{z_1^*} = f_1(0) + \Omega_1(q_1, \ldots q_K) = 0 \qquad (11.11)$$

so that $\Omega_1(q_2, \ldots q_K) = 0$ also and so, denoting q_1 as q_1, (11.10) becomes

$$z_1^* = f_1(q_1) \qquad (11.12)$$

and (11.8) becomes

$$u_1 = f_1(q_1) - z_1 \qquad (11.13)$$

which is expression (A11.2e) in the appendix.

11.2.2 Efficient Resource Allocation Within the Coalition

In economics and game theory, efficiency is identified with Pareto optimality. This in turn means that no-one can be made better off, within some objective material constraints, without shifting resources in such a way as to make another person worse off. Nevertheless, typically, there are an uncountable infinity of Pareto optima. Given measures of the benefits to each individual, u_i, and a mathematical representation of the constraints on these benefits, we may think of an efficient joint strategy or allocation of resources as one that corresponds to the maximum of a weighted sum of the benefits, $\Sigma \lambda_i u_i$, where $\Sigma \lambda_i = 1$ and the distribution weights λ_i express the relative importance of the distribution to agent i. For an idealized profit-maximizing firm, for example, $\lambda_i = 1$ for the proprietor and $\lambda_i = 0$ for everybody else, while for an idealized consumers' cooperative, λ_i has the same positive value for all customers and is zero for everybody else. Each distinct vector of *distributive weights* λ_i may give rise to a distinct Pareto optimum, and that is why there is, in general, an uncountable infinity of them.

For a model of the business firm, the objective constraints are the effective supply of labor and its productivity, along with the outside options or rationality constraints for the individual members. If we ignore the idiosyncratic productivity differences among employees, then the efficient policies of the firm are quite familiar from established economic theory. Effort is allocated among the employees so that each has the same marginal disutility of labor (this is equation A11.7j). This marginal disutility of labor, expressed in money terms, defines the marginal cost of production (equation A11.7m). Output is allocated among the customers in such a way that each has the same marginal utility of the good or service (equation A11.7h) and output is expanded until, again in money terms, the marginal utility of the good or service is equal to marginal cost (equation A11.7m). We may then identify marginal cost with the quotient of the marginal disutility of effort (common to all employees) and the marginal productivity of labor:

$$MC = \frac{\dfrac{\partial g_i}{\partial h_i}}{\dfrac{\partial F}{\partial L}} \tag{11.14}$$

This marginal cost is the efficient price p_1 in equations (11.5)–(11.7) above. Thus, each customer will take a quantity q_1 of the coalition's good or service such as to make the customer's marginal utility equal to this efficient marginal cost price.

When we allow for idiosyncratic differences in productivity, the conditions are a little more complex and less familiar. Once again output will be allocated so that all customers have the same marginal utility for the product or service of the coalition. We may define the value marginal product of labor (VMP) as the marginal product of aggregate labor times the common marginal utility of the good or service. The effort commitment of agent i will then be such that i's marginal disutility of labor is equal to the product of the idiosyncratic productivity k_i and the value marginal product of aggregate labor (equation A11.7k). Thus, the more able will be assigned to work somewhat longer hours or given more difficult assignments, given the same disutility-of-effort function. This is a result of the linear aggregation of abstract labor from effort commitments by heterogeneous employees: a different aggregation scheme might give different results.

In this case the determination of marginal cost is again a little more complex. In place of (11.14) we have

$$MC = \frac{\dfrac{\partial g_i}{\partial h_i}}{\dfrac{\partial F}{\partial L}}\frac{1}{k_i} \tag{11.15}$$

(equation A11.7n). Since $\frac{\frac{\partial g_i}{\partial h_i}}{k_i}$ is the same for all employees (equation A11.7g) expression (11.15) is well defined. Once again, the efficient price p_1 is the marginal cost price.

11.2.3 Independence of Production and Distribution

Thus, the quantity produced, its distribution among the customers, and the labor effort supplied and its distribution among the employees are all determined by the condition of Pareto optimality, and are identical for any set of distributive weights and so for any distribution of the benefits net of side payments. Since the normalization allows us to interpret the total benefit as the sum of the producers' and consumers' surpluses, and this is determined by the efficient allocation of effort, efficient total production, and efficient allocation of output among the customers, we may identify this sum as the gross value produced by the coalition, V^* in equations (A11.8a) and (A11.8b). The net surplus will then be that amount minus the sum of the rationality constraints of the members of the coalition, V in equation (A11.8b). This surplus will then be distributed according to bargaining power as shown in equations (A11.8f)–(A11.8j). Equations (A11.9a)–(A11.9m) demonstrate that the same conditions are derived when the two sets of decisions, allocative and distributive, are taken separately.

In summary, then, the coalition allocates its resources just as they would be allocated in a hypothetical perfectly competitive market in which the demand side comprises the customers and the supply side comprises the proprietor and employees. For such a perfectly competitive market, the equilibrium (marginal) price would correspond to equality between marginal cost and the money expression of the marginal utility to consumers, that is, to equality of supply and demand.

This is often presented as a proof that competition promotes efficiency. Further, we find that the perfectly competitive model often predicts developments in markets that divert far from the assumptions of "many small firms," and so on. This fact is sometimes represented as proof that competition is more pervasive than the model suggests. But this is all quite backward. What the value creation model tells us is that when (ideally) rational people engage in the cooperative activity of production for exchange, they will find ways to coordinate their action efficiently. A model that describes efficient allocations of resources is thus likely to be predictive of their actions. But it is cooperation, not competition, that leads to efficiency in market outcomes, to the extent that market outcomes are efficient. This calls in particular for reconsideration of models, such as monopoly and monopolistic competition, in which market outcomes are supposed to be inefficient in the absence of externalities, taxes or subsidies.

11.3   SOME POLICY ISSUES

The neoclassical economics of public policy assumes that exchange is cooperative, but that all other economic phenomena, including the terms of exchange, are determined by noncooperative decisions. If we conceive the firm as a cooperative coalition, we might expect that the implications for public policy would differ, in some cases. This section explores the implications, in this last-stage model, of some classical cases in the economics of public policy: monopoly power and its regulation, excise taxes and subsidies per unit of output.

11.3.1   Monopoly Reconsidered

The theory of monopoly in the late nineteenth and twentieth centuries has been little more than an elaboration of a passing remark of John Stuart Mill, who writes in the *Principles of Political Economy* (1865/1898, p. 272) "The monopolist can fix the price as high as he pleases, . . . but he can do so only by limiting the supply." This is true (and the Marshall–Lerner conditions make the point more precise; Lerner, 1934) provided the law of one price applies, so that price discrimination, all-or-nothing offers, and such can be excluded. Now, Mill was aware that the law of one price cannot always be applied where competition is imperfect (1865/1898, p. 149) but his argument here is that "Monopoly value . . . is but a mere variety of the ordinary case of demand and supply" (p. 272). But with the word "value" Mill adopted a law of one price and a long-run equilibrium perspective from classical value theory. For classical value theory value is a *unique* central tendency toward which prices, determined in the short run (p. 273) by supply and demand, must tend. This idea that a monopolist profits by restricting output has continued to dominate neoclassical economics throughout its history.

There are some firms that have monopoly power but have a "competitive fringe" and so cannot effectively make price-discriminatory offers. However, a literal monopoly will be able to make such offers, *and will always increase the surplus by doing so*. And increasing the surplus will increase the profit, along with the net benefit to other members of the coalition who possess positive bargaining power.

This applies with particular force to monopolistic competition. Where monopoly power is based on product differentiation there will not, in general, be a competitive fringe. "Each firm has a monopoly of its own product" (Robinson and Eatwell, 1974, p. 173) and there can be no fringe of competitive sellers of the firm's own brand. Monopolistic competition will indeed be the general case in a world of positive transaction costs, but for many purposes, we can apply the perfectly competitive model – marginal cost equal to marginal willingness to pay, supply equal to demand – to firms in such a monopolistically competitive situation, if we assume, as this chapter does, that prices are determined cooperatively.

An exception that deserves mention is the sale of tickets to performances, where the events are differentiated but the tickets can be "scalped" or resold. The scalpers are an instance of a competitive fringe. If a company that puts on such performances restricts its output in order to raise the price, it seems that it is competition, as much as monopoly, that causes the inefficient allocation of resources in such a case. Exceptions may also include cases in which a law of one price is imposed by law or regulation, that is, where price discrimination is prohibited. This may be a necessary condition for the allocation of resources through an auction mechanism, which is itself in the last analysis a cooperative arrangement (McCain, 2014a, pp. 65–6, 123–5). In some cases, however, regulations that exclude price discrimination may lead to an inefficient restriction of output.

But the key point is that monopoly power, *per se*, does not distort resource allocation away from efficiency. The impact of monopoly power is thus to be seen in the distributive, not the allocative decisions of the coalition for production and sale. Monopoly may well be, in Mill's words, "the taxation of the industrious for the support of indolence, if not of plunder" (1865/1898 p. 476), but not a source of inefficient resource allocation. This is important, for present purposes, because antimonopoly policy is generally premised on the supposed inefficiency of resource allocation under unregulated monopoly. It would seem that the whole discussion of economic policy toward monopoly, in the twentieth century, has been based on a mistaken premise. The impact of monopoly power is to be found, instead, in the bargaining power of the monopoly *vis-à-vis* its customers, and thus in the realm of distribution.

In the context of monopolistic competition, monopoly power may have little influence even on bargaining power. Monopolistic competition may take place among quite small businesses with products differentiated by location, reputation, or style. For an industry of this kind, nothing seems to be lost if we apply the perfectly competitive model, in that it will predict the *marginal* price and the quantity sold accurately. In short, monopolistic competition *per se* has no prima facie policy implications distinct from those of pure or perfect competition.

Thus far we have discussed decisions by an individual monopoly or monopolistically competitive coalition. A case of oligopoly – that is, where interactive noncooperative decisions of different firms are important – may yield different results.

11.3.2   Regulation of Monopoly Price

For larger-scale monopolies, the monopolist's bargaining power may have relevance for public policy. Historically, the regulation of monopoly in the United States gained momentum as a response to the bargaining power gained by railroads as against local communities that were served by only one line. This led to price discrimination that was widely viewed as unfair (Gifford and Kudrle, 2010, p. 1261). Indeed, this was the predominant understanding of regulation among lawyers and the general public (but not neoclassical economics) until the last quarter of the twentieth century (ibid., pp. 1238, 1255–6). Regulation was to a considerable extent the regulation of price discrimination, and deregulation was the deregulation of price discrimination. An argument for this deregulation of price discrimination begins by "stressing the pervasiveness of bargaining in intermediate good markets and the consequently misleading models that ignore this reality." Indeed! But the argument continues "forbidding price discrimination 'constrains the bargaining process by inhibiting buyers from seeking marginal price concessions that lower retail prices'" (Gifford and Kudrle, 2010, p. 1253, quoting O'Brien and Shaffer, 1994).

All of this, as stated, is consistent with the model in this chapter. It is, however, stated in a rather extreme form. It seems that one purpose of the regulatory laws on monopoly was indeed to "constrain the bargaining process" in those cases when it would lead to *unfair* price discrimination. Once we realize that rational bargainers will not agree on an inefficient price-and-sales strategy, this constraint may achieve a (normative) distributional adjustment without consequent inefficiency.

To model the regulation in terms of two-part tariffs as in the appendix to this chapter, we would add to equation (A11.4.b) or (A11.9f) multiplier terms for a set of upper constraints on customer side payments of the form

$$z_\iota \leq z_\iota^0 \; \forall \iota \in B \tag{11.16}$$

where B is the set of all customers of the firm. Then z_ι^0 might be a constant for all ι or might be exogenously determined by a rule of equal treatment of equals; for example z_ι^0 might be a fixed proportion of income. The additional multiplier term (for equation A11.4b) would be

$$\sum_{\iota \in B} \mu_{\iota,6}(z_\iota^0 - z_\iota) \tag{11.17}$$

The constraint may be binding in some cases and not in others, depending on the individual characteristics of the customer. For example, we may think of a customer who is linked to no other feasible coalitions that could supply a close substitute good (perhaps because she is not located at a rail junction) as against another customer who has other links (located at a rail junction, perhaps) and thus alternative sources of supply. The monopolist coalition would have greater monopoly power in the former case than in the latter, so that, plausibly, the former customer would have less bargaining power, β_i, than the latter. We might then find that (11.17) binds for the former but not in the latter. Let C be the set of customers for which it is binding.

Ideally, then, (11.17) would be binding in cases of "unfair" monopoly prices but not otherwise. In cases in which it is not binding, $\mu_{i,6} = 0$. In either case, equation (A11.6d) in the appendix to this chapter is replaced by

$$\forall\iota \in C, y_\iota = \frac{\beta_\iota}{\mu_4 - \mu_6} \tag{11.18}$$

$$\forall\iota \in B|C, y_\iota = \frac{\beta_\iota}{\mu_4} \tag{11.19}$$

Other necessary conditions would be unchanged. All in all, it would seem that, for most of the twentieth century, the legal framework for understanding the regulation of monopoly was sounder than that of the (neoclassical) economists. More recently, the economists' views having prevailed, it seems that all the professionals have it wrong.

It must again be stressed that all of this applies to the decisions of a given coalition in a given coalition structure, that is, a coalition with given membership. In later chapters, when the membership of the firm is variable, these points will need to be revisited.

11.3.3 The Incidence of Excise Taxes and Subsidies

If we consider a group of monopolistically competitive firms, many policy issues can be resolved by the application of a "perfectly competitive" model, that is, one based on equality of marginal cost and the marginal willingness to pay. Consider, for example, the classical issue of the incidence of an excise tax. In the appendix, we may modify equation (A11.8a) to read

$$V = \sum_{\iota \in B} f_\iota(q_\iota) - \sum_{i \in E} g_i(h_i) - t \sum_{\iota \in B} q_\iota \tag{11.20}$$

where t is the tax rate. The necessary conditions will then reproduce those for a market equilibrium with the same marginal willingness to pay and marginal cost functions, with a result that depends in a familiar way on the elasticities of the two functions. A numerical example may be found at McCain (2014b, pp. 425–8). The same reasoning will be applied to subsidies.

What seems of most interest here is the contrast with the previous subsection. Where monopoly and regulations based on monopoly power are addressed, the conclusions contrast sharply with those from neoclassical economics. By contrast, the conclusions of neoclassical microeconomics with respect to the incidence of excise taxes and production subsidies are not only confirmed but extended to cases in which the neoclassical analysis is not applicable, at least in its simplest form. The lesson seems to be that there is no substitute for the reconsideration of each specific point of microeconomics on the basis of the cooperative conception of the firm. There can be no wholesale rejection nor confirmation of the orthodox analysis. Rather the results are likely to differ from case to case.

11.4 SOME LIMITATIONS OF THE MODEL

Every economic model makes simplifying assumptions, and the simplifying assumptions set limits on the applicability of the model. That the model in this chapter only considers the last, cooperative stage of the coalition formation and decisions is a limitation that will be reconsidered in later chapters.

The model presented here is strictly a rational-action, full-information model. The only informational limitation assumed is that cooperative decisions can only be made among individuals who are linked in a previously-formed network, and this is not relevant to this chapter, which takes the coalition as formed. However, if we recognize that human rationality is bounded and is commonly expressed in "optimally imperfect" rules of thumb (Baumol and Quandt, 1964; Simon 1978; McCain 2014a, Chapters 6–7) then the maximization model presented here can only be approximated.

The normalizations discussed in section 11.2.1 lead to the independence of allocative decisions and distributive decisions. Conventional as they are, they are far-reaching. If we apply the normalization of equation (11.9) to all products simultaneously, we are assuming that utility is additively separable in all products without exception. But this is almost certainly empirically false. Empirical realism would be served if instead we adopted the non-transferable utility approach along the lines of McCain (2013, Chapter 6, section 5), but for that model the numerical expression of the firm's value created is measured in units unique to that firm. Conversely, when we measure the output of a firm as its value added, and total output as GDP, we are implicitly assuming additive separability, whether we acknowledge it or not. Perhaps additive separability is best considered as a local approximation to a more complex global reality.

11.5 SUMMARY

This chapter has explored the decisions of a cooperative coalition for production and exchange, with given membership, as the last stage in a multi-stage decision process in a context of backward induction. While the results of this discussion can only be tentative, awaiting analysis of the earlier stages of the decision process, they suggest that the impacts of public policy will be quite different, in some cases, than predicted by neoclassical economics. For monopoly theory, for example, the cooperative approach predicts no impact on the allocation of resources (at least at this stage) and consequently would call for a reconsideration of the regulation of monopoly. For excise taxes and subsidies, by contrast, the cooperative approach confirms the analysis of the neoclassical perfectly competitive model, and indeed extends it to monopolistic competition and monopoly. What all of this seems to show is that each proposition in the economic theory of public policy will require a specific reconsideration if we adopt a cooperative approach to the theory of the firm.

NOTES

1. At the more advanced research level, especially in the microfoundations of macroeconomics, the firm is often represented as a producer of a differentiated intermediate good that is transformed by a "black box" competitive mechanism to an undifferentiated consumer good sold on perfectly competitive terms. This allows a reconciliation between product differentiation and a relatively simple and tractable model of saving and investment. This book will argue instead for a return to a "perfectly competitive" model, although based on quite different assumptions than the traditional competitive model.
2. Recall, at an earlier stage the employees have been recruited via a costly search-and-matching process. The variation in idiosyncratic productivity, depending on the match between the employee and the job, motivates this costly search. This will be explicitly considered in a later chapter.
3. For a discussion of the range of possibilities at the level of intermediate microeconomics and in a thoroughly noncooperative context, see Goosbee et al. (2013, Chapter 10, pp. 395–433).

APPENDIX A11

The coalition for production and sale comprises three subgroups. The membership of the coalition is $N = A \cup B \cup E$, where A comprises the proprietors, B the customers, and E the employees. For simplicity we suppose that there is a single proprietor. The relationship between inputs and output in production is described by three constraints. The first determines the labor resource available to the firm.

$$L \le \sum_{i \in E} k_i h_i - L_0 \tag{A11.1a}$$

The second is a conventional "short run" production function.

$$Q \le F(L) \tag{A11.1b}$$

The third explicitly limits sales to production in the period.

$$\sum_{i \in B} q_i \le Q \tag{A11.1c}$$

$$\forall i \in E, u_i = z_i - g_i(h_i) \tag{A11.2a}$$

where u_i is the utility payoff to employee i, z_i the money payoff, h_i effort, and $g_i(.)$ the disutility of effort. The function g is assumed to have continuous first and second derivatives.

$$g(h_i) \ge 0 \tag{A11.2b}$$

$$\frac{\partial g_i}{\partial h_i} > 0 \text{ and } \frac{\partial g_i^2}{\partial^2 h_i} > 0 \text{ if } h_i > H_i \ge 0 \tag{A11.2c}$$

$$\forall i \in B, u_i = f_i(q_i) - z_i \tag{A11.2d}$$

where u_i is the utility payoff to customer i, q_i the allocation of the coalition's product or services to that customer, $f_i(q_i)$ the utility of that allocation, and z_i the customer's money side payment into the coalition. For every $i \in N$, denote by w_i the outside option for agent i and by y_i the utility payoff in excess of the outside option. Then

$$\forall i \in E, y_i = z_i - g_i(h_i) - w_i \tag{A11.3a}$$

$$\forall \iota \in B, \, y_\iota = f_\iota(q_\iota) - z_\iota - w_\iota \qquad \text{(A11.3b)}$$

$$y_A = \Pi - w_A \qquad \text{(A11.3c)}$$

$$\Pi = \sum_{\iota \in B} z_\iota - \sum_{i \in E} z_i \qquad \text{(A11.3d)}$$

We will initially consider the determination of the optimal policy for the coalition, including the assignment of h_i and q_ι and the imputation of the surplus as payments to the members as simultaneous. For the bargaining model used here, the imputation of the benefit to the individual members depends on their relative gains over their outside options. However, the benefits and the outside options are expressed in utility terms, and thus the gains cannot be compared without distributional weights. The distributional weights, λ_i, will themselves be determined by the bargaining process (see McCain, 2013, Chapter 6, section 6.5). However, distributional weights are determinate only up to a multiplicative transformation. Thus, to make the weights determinate, we will require a normalizing constraint:

$$\sum_{i \in N} \lambda_i \leq 1 \qquad \text{(A11.3e)}$$

Thus we have the objective function

$$y = \sum_{\iota \in B} \beta_\iota \ln(\lambda_\iota y_\iota) + \sum_{i \in E} \beta_i \ln(\lambda_i y_i) + \beta_A \ln(\lambda_A y_A) \qquad \text{(A11.4a)}$$

where β_j is an index of the individual bargaining power of individual j. The Lagrangean function is

$$\mathcal{L} = y + \mu_1 \left(\sum_{i \in E} h_i k_i - L_0 - L \right) + \mu_2(F(L) - Q) + \mu_3 \left(Q - \sum_{\iota \in B} q_\iota \right)$$

$$+ \mu_4 \left(\sum_{\iota \in B} z_\iota - \sum_{i \in E} z_i - \Pi \right) + \mu_5 \left(1 - \sum_{i \in N} \lambda_i \right) \qquad \text{(A11.4b)}$$

Now consider the determination of the distributive weights.

$$\forall \iota \in B, \frac{\partial \mathcal{L}}{\partial \lambda_\iota} = \frac{\beta_\iota}{\lambda_\iota} - \mu_5 \leq 0 \qquad \text{(A11.5a)}$$

$$\forall i \in E, \frac{\partial \mathcal{L}}{\partial \lambda_i} = \frac{\beta_i}{\lambda_i} - \mu_5 \leq 0 \qquad \text{(A11.5b)}$$

$$\frac{\partial \mathcal{L}}{\partial \lambda_A} = \frac{\beta_A}{\lambda_A} - \mu_5 \leq 0 \qquad\qquad \text{(A11.5c)}$$

Thus we have

$$\forall i \in N, \lambda_i = \frac{\beta_i}{\displaystyle\sum_{j \in N} \beta_j} \qquad\qquad \text{(A11.5d)}$$

That is, for each member of the coalition, the distributive weight is the same as the relative bargaining power. This may seem a common-sense result, but in fact it is to be expected only in a TU game; moreover we have purchased it at the cost of a strong restriction: that only individuals have bargaining power. This will be reconsidered in a later chapter. Note that, in each case, an interior solution is necessary since a zero value for any λ_i would result in a value for $\mathcal{L}$ of minus infinity.

Now consider the necessary conditions for the individual money payments into and out of the coalition. We have

$$\forall i \in E, \frac{\partial \mathcal{L}}{\partial z_i} = \frac{\beta_i}{y_i} - \mu_4 \leq 0 \qquad\qquad \text{(A11.6a)}$$

$$\forall \iota \in B, \frac{\partial \mathcal{L}}{\partial z_\iota} = -\frac{\beta_\iota}{y_\iota} + \mu_4 \leq 0 \qquad\qquad \text{(A11.6b)}$$

$$\frac{\partial \mathcal{L}}{\partial z_A} = \frac{\beta_A}{y_A} - \mu_4 \leq 0 \qquad\qquad \text{(A11.6c)}$$

Again, interior solutions are necessary for a finite value for the objective function. Thus,

$$\forall j \in N, y_j = \frac{\beta_j}{\mu_4} \qquad\qquad \text{(A11.6d)}$$

That is, the net benefit from the coalition, in utility terms, is proportionate to the individual's bargaining power. Again, this common-sense point has been bought at the cost of the simplifying assumption that only individuals have nonzero indices of bargaining power. Conversely,

$$\forall j \in N, \frac{\beta_j}{y_j} = \mu_4 \qquad\qquad \text{(A11.6e)}$$

Now consider the conditions for the strategy of production and sale.

$$\frac{\partial \mathcal{L}}{\partial L} = -\mu_1 + \mu_2 \frac{\partial F}{\partial L} \leq 0 \qquad\qquad \text{(A11.7a)}$$

$$\frac{\partial L}{\partial Q} = -\mu_2 + \mu_3 \leq 0 \qquad \text{(A11.7b)}$$

In these cases, a corner solution would mean that production is shut down, and this contingency is excluded by assumption.

$$\forall i \in E, \frac{\partial \mathcal{L}}{\partial h_i} = -\frac{\beta_i}{y_i}\frac{\partial g_i}{\partial h_i} + \mu_1 k_i \leq 0 \qquad \text{(A11.7c)}$$

Note that, in this case, the inequality will be violated unless $\frac{\partial g_i}{\partial h_i} > 0$, and this will be true only for positive values of h_i, so that an interior solution is necessary. Thus we have

$$\forall i \in E, \frac{\beta_i}{y_i}\frac{\partial g_i}{\partial h_i} = \mu_1 k_i \qquad \text{(A11.7d)}$$

Further,

$$\forall \iota \in B, \frac{\partial \mathcal{L}}{\partial q_\iota} = \frac{\beta_\iota}{y_\iota}\frac{\partial f_\iota}{\partial q_\iota} - \mu_3 \leq 0 \qquad \text{(A11.7e)}$$

In this case, a corner solution cannot be ruled out by formal considerations, but nevertheless we will exclude a corner solution on the reasoning that, if a customer did not anticipate purchasing a positive amount, he or she would not join the coalition. Then

$$\forall \iota \in B, \frac{\beta_\iota}{y_\iota}\frac{\partial f_\iota}{\partial q_\iota} = \mu_3 \qquad \text{(A11.7f)}$$

Taking (A11.7d) and (A11.6e), we have

$$\forall i \in E, \frac{\partial g_i}{\partial h_i} = \frac{\mu_1}{\mu_4}k_i \qquad \text{(A11.7g)}$$

Similarly, taking (A11.7f) and (A11.6e),

$$\forall \iota \in B, \frac{\partial f_\iota}{\partial q_\iota} = \frac{\mu_3}{\mu_4} \qquad \text{(A11.7h)}$$

Equation (A11.7h) is a familiar efficiency condition, which tells us that the production of the coalition should be allocated among customers in such a way that their marginal utilities are not different. Equation (A11.7g) is less familiar because it takes account of idiosyncratic differences in productivity among employees. Suppose instead that $\forall i \in E, \forall j \in E, k_i = k_j = 1$. Then (A11.7g) becomes

$$\forall i \in E, \frac{\partial g_i}{\partial h_i} = \frac{\mu_1}{\mu_4} \qquad \text{(A11.7j)}$$

that is, effort is allocated among employees in such a way that all employees experience the same marginal disutility of effort. It is of interest that these conditions do not depend on the bargaining power or the distributive weights attributable to the individuals. They are conditions for Pareto efficiency per se.

Taking (A11.7g) and (A11.7h) together with (A11.7a) and (A11.7b),

$$\forall i \in E, \forall \iota \in B, \frac{\partial g_i}{\partial h_i} = \frac{\partial f_\iota}{\partial q_i} \frac{\partial F}{\partial L} k_i \qquad (A11.7k)$$

This generalizes another familiar efficiency condition, namely that the marginal value product of labor is set equal to the price of output. Denoting the shadow price of the firm's output as p_s, and again assuming $k_i = k_j = 1$, (A11.7k) becomes

$$\forall i \in E, \frac{\partial g_i}{\partial h_i} = p_s \frac{\partial f}{\partial L} \qquad (A11.7m)$$

Equation (A11.7k) is equivalent to

$$\frac{\partial f_\iota}{\partial q_\iota} = \frac{\frac{\partial g_i}{\partial h_i}}{\frac{\partial F}{\partial L}} \frac{1}{k_i} \qquad (A11.7n)$$

which may be read as "marginal utility equals marginal cost."

Now, let

$$V^* = \sum_{\iota \in B} f_\iota(q_\iota) - \sum_{i \in E} g_i(h_i) \qquad (A11.8a)$$

Here, V^* is the gross value created by the coalition expressed in money terms using the normalization assumptions we have adopted. That is, it is the sum of consumers' and producers' surplus for this example. Then let

$$V = V^* - \left(w_A + \sum_{\iota \in B} w_\iota + \sum_{i \in E} w_i \right) \qquad (A11.8b)$$

This is the net surplus over the outside options of the coalition members. We then have

$$\sum_{i \in N} y_i = V \qquad (A11.8c)$$

From (A11.6d),

$$\sum_{i \in N} y_i = \frac{1}{\mu_4} \sum_{i \in N} \beta_i \qquad (A11.8d)$$

Thus

$$\mu_4 = \frac{\sum\limits_{i \in N} \beta_i}{V} \qquad \text{(A11.8e)}$$

Then, again using (A11.6d) and also (A11.5d)

$$y_i = \frac{\beta_i}{\sum\limits_{j \in N} \beta_j} V = \lambda_i V \qquad \text{(A11.8f)}$$

Thus, using also (A11.3a),

$$\forall i \in E, z_i = g_i(h_i) + w_i + \lambda_i V \qquad \text{(A11.8g)}$$

The wage is the outside option plus the disutility of effort plus a proportion of the surplus. Using (A11.3b),

$$\forall \iota \in B, z_{\iota} = f_{\iota}(q_{\iota}) - w_i - \lambda_i V \qquad \text{(A11.8h)}$$

The customer's payment into the coalition is the utility from consuming the coalition's product or service minus an amount that comprises the outside option plus a proportion of the surplus. Using (A11.3c),

$$\Pi = w_A + \lambda_A V \qquad \text{(A11.8j)}$$

Profit is the outside option of the proprietor plus a proportion of the surplus that depends on the proprietor's relative bargaining power.

Recall that from (A11.7g), (A11.7h), and (A11.8a), the value created by the coalition is independent of the distributive multipliers and thus of the relative bargaining powers. Thus, tentatively, we might equivalently treat the decision process as a two-stage process in which the allocative decisions, determining L, Q, h_i, q_ι, and V, are made with the objective of maximizing V or of attaining Pareto optimality and the determination of the individual payoffs y_i is determined by bargaining in a TU game over V.

Suppose, then, that at the first stage the coalition maximizes V^* in expression (A11.8a) subject to constraints (11.1a), (11.1b), and (11.1c). We have the Lagrangean function

$$\mathcal{L}^* = V^* + \mu_1 \left(\sum_{i \in E} h_i k_i - L_0 - L \right) + \mu_2 (F(L) - Q)$$

$$+ \mu_3 \left(Q - \sum_{\iota \in B} q_\iota \right) \qquad \text{(A11.9a)}$$

Necessary conditions again include (A11.7a) and (A11.7b). Further necessary conditions are

$$\frac{\partial \mathcal{L}}{\partial h_i} = -\frac{\partial g_i}{\partial h_i} + \mu_1 k_i \leq 0 \qquad \text{(A11.9b)}$$

$$\frac{\partial \mathcal{L}}{\partial q_1} = \frac{\partial f_1}{\partial q_1} + \mu_3 \leq 0 \qquad \text{(A11.9c)}$$

Again assuming interior solutions, this will yield (A11.7g), (A11.7h), and (A11.7k). Thus the maximization of V^* will lead to the same values of L, Q, h_i, q_1, and V^* as the NTU bargaining model does.

Now suppose that the coalition simply distributes V according to bargaining power. That is,

$$max\, Y = \sum_{i \in N} \beta_i \ln(y_i) \qquad \text{(A11.9d)}$$

subject to

$$\sum_{i \in N} y_i \leq V^* - \sum_{i \in N} w_i = V \qquad \text{(A11.9e)}$$

The Lagrangean function will be

$$\mathcal{L}^{**} = Y + \mu \left(V^* - \sum_{i \in N} y_i - \sum_{i \in N} w_i \right) \qquad \text{(A11.9f)}$$

A necessary condition for the maximum is

$$\frac{\partial \mathcal{L}}{\partial y_i} = \frac{\beta_i}{y_i} - \mu \leq 0 \qquad \text{(A11.9g)}$$

As before, an interior solution is the only possibility. Thus,

$$y_i = \frac{\beta_i}{\mu}, \qquad \text{(A11.9h)}$$

which yields, as before,

$$y_i = \frac{\beta_i}{\sum_{j \in N} \beta_j} V, \qquad \text{(A11.9j)}$$

that is,

$$\forall i \in E,\, z_i = \frac{\beta_i}{\sum_{j \in N} \beta_j} V + g_i(h_i) + w_i \qquad \text{(A11.9k)}$$

$$\forall \iota \in B, z_\iota = f_\iota(q_\iota) - \frac{\beta_\iota}{\displaystyle\sum_{j \in N} \beta_j} V - w_\iota \qquad \text{(A11.9m)}$$

Thus, it seems that we may indeed treat the decisions of the firm as a TU game, with maximization of the money value of the firm's activities at the first stage and distribution according to bargaining power at the second stage.

12. What coalitions will be formed?

Chapter 11 explored the cooperative decisions of an existing coalition for production and exchange. Following the logic of backward induction, this chapter addresses the prior question: what coalitions will be formed? For the purposes of this part of the book, that question has two parts: (1) within a given set of information-sharing links, some feasible coalitions may nevertheless be inconsistent. For example, a particular agent cannot both be included and excluded from a particular coalition. Thus, the first question is: which of these coalitions will form? (2) What will influence or determine the links that are formed and maintained? In addressing the first question we will again draw on ideas from cooperative game theory, in the broad tradition of the theory of the core; for the second, noncooperative game theory and random processes.

12.1 RECONTRACTING

In the theory of superadditive games in coalition function form, the *core* of the game comprises coalitions and payoff schedules that are stable under a process of recontracting. (Chapter 8 above). It is not clear that recontracting is "cooperative" in the fullest sense, since each individual or group chooses which of two or more coalitions to affiliate with on the basis of the benefits they receive, regardless of the benefits to others. Nevertheless, (1) it is presupposed that, once the recontracting process is finished, coalitional decisions will be for mutual benefit. Perhaps recontracting could be more clearly called "contract shopping." (2) Recontracting captures important aspects of "competition" as neoclassical economics uses the term.

Economics uses the term "competition" in a special and limited sense. Consider, as an example that contrasts with the economist's usage, a neighborhood loan-shark who responds to competitive entry by having the competitor's legs broken. In the non-economist's use of the word "competitive," the loan-shark's action is extremely and aggressively competitive! However, the economic theory of competition assumes away any possibility of competitive threats of this sort. Similarly, a government-connected oligarch who has his competitor jailed for alleged tax evasion is not acting "competitively" as economic theory uses the term. Economic theory allows only one sort of competitive strategy: making a better offer. In simpler economic theories a "better offer" means a cheaper price, although we recognize that in practice the offer may have nonprice dimensions as well. Perhaps it would be clearer if the phrase "competitive offering" were substituted for "competition." In any case, the recontracting process generalizes competitive offering, as the literature on the core of a market game demonstrates (Chapter 8, section 8.2.1 above). Accordingly, stability under recontracting is extended to the class of games we now consider.

12.1.1 The Core

For this chapter, a coalition is *feasible* if its members are linked in appropriate ways. McCain (2013, p. 193) suggests that such a group will comprise a central agent, along with agents directly linked to that central agent. More complex definitions of "appropriate linkage" might be suggested (McCain, 2013, also briefly mentions a k-degrees-of-separation criterion as an alternative) but, for simplicity, we will adopt the former, single-link criterion. A feasible coalition structure (FCS) is a set of appropriately linked coalitions that may meet some other criteria as well. For this chapter, the coalition structure will not be feasible if certain kinds of members, including employees, proprietors, and the central agent to whom all are linked, are duplicated in more than one coalition.

For this chapter we consider an FCS game in which considerable solutions are determined as in Chapter 11. A candidate solution is a (feasible) coalition structure $Q = \{S_1, S_2, \ldots\}$ and a schedule of net payoffs $\{y_i\}$ from the coalitions to their members, with $\sum_{i \in S_j} y_i = V(Q, S_j)$, where $V(Q, S_j)$ is the value created by coalition S_j if the coalition structure Q is formed. In the context of considerable solutions, the value of C will depend on the coalition structure Q in which C is imbedded via the Nash equilibrium among the separate coalitions. Suppose then that a deviation D is proposed. That is, D is a coalition among agents who are appropriately linked but $D \notin Q$. Let $\sigma(Q, D)$ be the rational successor function[1] for the FCS game formed by the existing information-sharing links. If $R = \sigma(Q, D)$, $D \in R$ and $V(R, D) > \sum_{i \in D} y_i$, then Q is *dominated* by R via D. That is, a candidate solution is dominated if a group of agents can form a new coalition D, forming a new coalition structure, which realizes more value among them than they derive from the existing coalition structure. The set of undominated candidate solutions comprises the *core* of the game. If for Q there is a payoff schedule $\{y_i\}$ such that the candidate solution $\{Q, \{y_i\}\}$ is undominated, then Q is *stable*.

Consider the following example. There are seven agents with links as

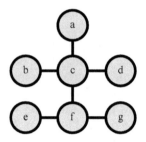

Figure 12.1   Links among seven agents

Table 12.1   Some feasible coalition structures

I	1. {a,b,c,d}(16) 2. {e,f,g}(9)
II	1. {a,b,c,d,f}(25) 2. {e}(1) 3. {g}(1)
III	1. {a}(1) 2. {b}(1) 3. {d}(1) 4. {c,e,f,g}(16)
IV	1. {a}(1) 2. {b}(1) 3. {c,d}(4) 4. {e,f,g}(9)

shown by the lines connecting nodes in Figure 12.1. For this example, coalitions can be formed only among agents linked to a common central agent, and no agent can be a member of more than one coalition. (These are links of the employee type, not of the customer type.) Assume further that the value of a coalition is the square of the number of agents in the coalition.

Table 12.1 shows four of a larger number of coalition structures feasible for this example, with coalitions indicated by brackets, {}, and the values of the coalitions following in parentheses.

The first three will be of primary interest, since they cannot be consolidated. By contrast, structure IV can be consolidated by merging {a}, {b}, and {c, d} producing structure I. Moreover, clearly, structure I and payouts of 4,4,4,4 for coalition I.1 dominates structure IV with any payoff schedule feasible for IV. In general, any structure comprising proper subsets of coalitions I, II, or III will be dominated (in this example) and thus not stable. Thus, we limit attention from this point to I, II, and III.

We will find that II is the only stable coalition structure. For II to be stable against I we require

$$y_a + y_b + y_c + y_d \geq 16 \qquad (12.1)$$

$$y_e + y_f + y_g \geq 9 \qquad (12.2)$$

and for stability against III,

$$y_c + y_e + y_f + y_g \geq 16 \qquad (12.3)$$

For feasibility

$$y_a + y_b + y_c + y_d + y_f \leq 25, y_e = 1, y_g = 1 \qquad (12.4)$$

Consider a payoff schedule

$$y_a = y_b = y_d = 2.33, y_c = y_f = 9 \qquad (12.5)$$

Inequalities (12.1)–(12.4) are satisfied, so the candidate solution comprising II with payoff schedule (12.5) is undominated and II is a stable coalition.

Similar reasoning will establish that I and III are not stable; and we have already noted that there are no other stable coalition structures in the example. Thus, for this example, recontracting specifies a unique stable coalition structure. (That will not be true in general without some powerful additional assumptions.) However, the payoff schedule is not uniquely specified. Consider the payoff schedule

$$y_a = y_b = y_d = 1.66, y_c = 12, y_f = 8 \qquad (12.6)$$

This payoff schedule is also undominated. In this example, as we assumed in the previous chapter, the payoffs will be determined in part by bargaining power.

12.1.2 Bargaining Power

Suppose, however, that (f believes) c has bargaining power of 0.8, while all other parties (including f) have equal bargaining power so that, in coalition II.1, the total bargaining power is 1. In coalition II.1 then, the payoff schedule would be $y_a = y_b = y_d = y_f = 1.25$, $y_c = 20$. In coalition I.2, by contrast, the payoffs implied by equal bargaining power would be $y_e = y_f = y_g = 3$. Thus, if f were to shift from I.2 to II.1, he (believes he) would be worse off. Accordingly, he would refuse to shift and (supposing f reasons in this way) coalition structure I is stable.

This is an instance of what I have elsewhere (McCain, 2013, pp. 94, 122) called the extended core. Should this somewhat relaxed concept of stability be chosen rather than the conventional concept of domination discussed in the previous subsection? There is, of course, no mathematically conclusive answer to this question. Only plausible arguments can be given. (Compare, for example, von Neumann and Morgenstern, 1944/2004, p. 506.) In that framework, the answer may depend on our understanding of bargaining power and its determinants. There is no widely received theory here! If we think of bargaining power as intrinsic to the person – superior ability to dissemble and bluff, for example, or optimism (as in Brandenburger and Stuart, 2007, p. 542) – then the extended core may seem to be the appropriate concept. If we think of bargaining power as something that arises from the coalition and its situation as a whole, then we might doubt that one person's bargaining power could be so great as to prevent the formation of a coalition that would otherwise be mutually beneficial.

Recall that bargaining power arises from threats, in that the other agents make concessions to prevent the carrying out of the threat. This will depend on the credibility of the threat. In coalition II.1, for example, c might threaten to withdraw, reducing all payoffs to 1. In a noncooperative framework, this threat would not be credible, but here we are assuming ideal rationality, and ideally rational agents will make such commitments if they expect to be better off as a result. But the commitment that makes c better off in this case will be a commitment NOT to withdraw. That is c, if rational, will make commitments that will voluntarily reduce his bargaining power so as to make the coalition II.1 possible.

On the basis of this plausible argument, the rest of this section will assume that a stable coalition structure will be one that is undominated, using the concept of domination as adapted in the previous subsection to FCS games. While this formulation does not support sweeping theorems on the uniqueness of pay schedules and similar matters, it does allow us to recover and somewhat extend some familiar ideas from neoclassical economics.

12.1.3 Marginal Productivity

The idea that an employee will be paid no more than her marginal productivity supplies an instance. Let $S \in Q$ be a coalition for production and sale, and let i be an employee. Consider the deviation $D = S\backslash\{i\}$. We assume that D is feasible, which is to say that the departure of $\{i\}$ does not disrupt links that would be necessary for $S\backslash\{i\}$ to function cooperatively. (That is, i is not a central agent in the links that permit S. In more ordinary terms, if, for example, i is a key salesperson who would take some customers with her, then the case will be a little more complex.) Let $R = \sigma(Q, D)$ be the rational successor to the separation of $\{i\}$ from S. (For example, i's second-best alternative might be to affiliate with another firm to which he is appropriately linked.)

The rationality constraint for D is

$$\sum_{\substack{j \in S \\ j \neq i}} y_j \geq V(\mathcal{R},D) \tag{12.7}$$

Then with

$$\sum_{j \in S} y_j = V(Q,S) \tag{12.8}$$

this implies

$$y_i \leq V(Q,S) - V(\mathcal{R},D) \tag{12.9}$$

Now, $V(Q,S) - V(\mathcal{R},D)$ is precisely employee i's contribution to the value created by the coalition. We conclude that no-one will be employed at a rate of pay that exceeds her or his marginal value product, and this requires us to assume neither homogeneity of labor nor profit maximization. (Compare McCain, 2013, pp. 212–14.)

But this requires a little further discussion. Recall that y_i is net of the employee's outside option and the compensation for the disutility of efficient effort, that is

$$y_i = z_i - g_i(h_i) - w_i \tag{12.10}$$

where z_i is the side payment to employee i, that is, the wage paid to i for the period and $g_i(h_i)$ and w_i are already deducted from $V(S)$. We would parallel neoclassical economics more closely if we were to consider the gross value added by coalition S as

$$V^*(Q,S) = V(Q,S) + \sum_{j \in S} g_j(h_j) + \sum_{j \in S} w_j \tag{12.11a}$$

Then the corresponding value of D is

$$V^*(\mathcal{R},D) = V(\mathcal{R},D) + \sum_{\substack{j \in S \\ j \neq i}} g_j(h_j) + \sum_{\substack{j \in S \\ j \neq i}} w_j \tag{12.11b}$$

Taking these together with (12.9), we have

$$z_i \leq V^*(Q, S) - V^*(\mathcal{R}, D) \tag{12.12}$$

and $V^*(Q, S) - V^*(\mathcal{R}, D)$ is the value marginal product of the individual employee i as more conventionally defined in neoclassical economics.

Taking this with the previous chapter we may say that the wage payment to an individual employee is bounded below by $g_i(h_i) + w_i$ and bounded above by $V^*(Q, S) - V^*(\mathcal{R}, D)$, the individual value marginal product; and if $g_i(h_i) + w_i > V^*(Q, S) - V^*(\mathcal{R}, D)$, the coalition encompassing employee i is unstable, so that the employee will be dismissed. And further note that this is no less applicable to a consumers' or worker cooperative than to a profit-maximizing firm.

12.1.4 Profit

Now consider the proprietor of the firm. Nothing has yet been said about the proprietor except that she provides the nonhuman means of production. At this level of abstraction the proprietor in effect is *identified with* a given set of nonhuman means of production, that is, capital goods, tracts of land, and stores of raw and semifinished materials. Thus, let proprietor $a \in S$, and a' another potential proprietor, and consider the deviation $D = (S \backslash \{a\}) \cup \{a'\}$. To say that a' is a potential proprietor is to say that a' is appropriately linked to the other members of S as a is. In this we assume that the rational successor $\mathcal{R} = \sigma(Q,D)$ is the naïve successor;[2] that is, we abstract from any reorganization of N/D. Suppose then a supplies nonhuman means of production $\{\alpha^*_1, \alpha^*_2, \alpha^*_3, \ldots\}$ and a' would supply a different set $\{\alpha'_1, \alpha'_2, \alpha'_3, \ldots\}$. We extend the production function, equation (A11.1b) in the appendix to Chapter 11, as

$$Q \le F(L, \{\alpha_1, \alpha_2, \alpha_3, \ldots\}) \tag{12.13}$$

With

$$F(L, \{\alpha^*_1, \alpha^*_2, \alpha^*_3, \ldots\}) < F(L, \{\alpha'_1, \alpha'_2, \alpha'_3, \ldots\}) \tag{12.14}$$

This will yield, for the same total effort, a greater Q and thus a greater total consumers' surplus than is attainable for coalition S; so that $V(\mathcal{R}, D) > V(Q, S)$. Since, however,

$$\sum_{i \in S} y_i \le V(S) < V(D) \sum_{i \in S} y_i \le V(Q,S) < V(\mathcal{R},D) \tag{12.15}$$

coalition structure Q is dominated via D and consequently Q is unstable.

To relate this to a capitalist economy will require some interpretation! Of course, it is incomplete to identify the proprietor with a bundle of capital goods. The proprietor will have determined, by an earlier, noncooperative decision, what nonhuman inputs to procure. What we have learned is that, if the proprietor has not procured the set of nonhuman inputs that generates maximum value for the coalition, then the coalition may be unstable. If for some reason the proprietor is unable to procure this optimal set of nonhuman inputs – because, for one example, of liquidity constraint and a consequent inability to raise the necessary finance or, for a second example, because land is entailed and cannot be bought and sold – then it may be profitable, and efficiency probably would be increased, if the firm were sold to a different proprietor.

In this discussion of the proprietor, we have so far ignored transaction costs, but the transition of proprietors will probably often entail very large transaction costs. For the model in this part of the book, transaction costs are the costs of search and matching to establish new information-sharing links, and what we have assumed is that the potential new proprietor is already appropriately linked to the coalition. If the "potential new proprietor" is the same person equipped with a different set of capital goods, this will be no difficulty. In the case of finding a new buyer for an undercapitalized firm, transaction costs could be quite large and a crucial consideration.

(There is also the complication that the proprietor might be a central agent, that is, c or f in Figure 12.1. In that case, however, the proprietor would not be a pure supplier of nonhuman inputs as assumed, but would also supply "human capital" in the form of links – goodwill – and thus is also an employee with an idiosyncratic productivity k_a that depends on that human capital. In small firms this will be the usual situation. However, to maintain the analytical distinction between the proprietor and employees, which is borrowed from neoclassical economics, the following convention is adopted: the central agent, who maintains information-sharing linkages among the members of the coalition, is an employee, the "human resources director," and is analytically distinct from the proprietor. In any case our examples of linkage are highly simplified, and the distinct "central agent" is herself a fiction of simplification.)

Now, suppose again that $a \in S$, and $D = (S \backslash \{a\}) < \{a'\}$, where a' is a potential proprietor appropriately linked to the members of S other than a. However, suppose instead that $\alpha^*_k = \alpha'_k$, that is, both supply the same set of nonhuman inputs. Then, by reasoning parallel to that associated with equations (12.13)–(12.15),

$$y_a \leq V(Q,S) - V(R,D) = 0 \qquad (12.16a)$$

Since

$$y_a = \Pi - w_a \qquad (12.16b)$$

in a stable coalition, the (economic or net) profit accruing to the proprietor is zero. In particular, suppose the money value of $\{\alpha^*_1, \alpha^*_2, \alpha^*_3, \ldots\}$

is K, the alternative rate of return is r; then $\Pi = rK$. This corresponds to the familiar neoclassical proposition that in a fully competitive situation, the rate of "economic profit" is zero.

Here again we have ignored transaction costs and simply assumed that an alternative proprietor with the same suite of nonhuman inputs is freely linked to the coalition. This seems a reasonable construction of a "fully competitive situation." Still, to illustrate the importance of this assumption, suppose the proprietor, a, is engaged in an innovation in the Austrian framework described by Schumpeter (1934/1961, pp. 65–6): a "new combination" of higher order goods. Then there is no other potential proprietor, linked or unlinked, who would bring $\alpha'_k = \alpha^*_k$ to production. Suppose $\{a'_k\}$ are the higher-order goods (nonhuman means of production) used within Schumpeter's (1934/1961, Chapter 1) circular flow. As they are routine, we may suppose without much hesitation that there will be an appropriately linked potential new proprietor who would supply $\{a'_k\}$ for deviation D. Thus

$$y_a \le V(S) - V(D) = \Pi - w_a \qquad (12.17a)$$

$$\Pi \le w_a + V(S) - V(D) \qquad (12.17b)$$

We see that, as Schumpeter argued, economic profit may be associated with innovations. (However, they will also depend on the proprietor's bargaining power.)

Two further remarks are needed. First, in practice, an innovator may need to create new links and establish a new coalition. This, too, is a "new combination" of higher-order goods (since labor is also a higher-order good) and the formation of a new coalition to use methods that are already well established is no less entrepreneurial, in the Austrian sense, than "the Deerfoot Sausage" (Schumpeter, 1947, p. 141). But this goes beyond the function we have conventionally associated with a pure proprietor, as, indeed, entrepreneurial activity often will. Second, we cannot concede to Schumpeter that *only* innovation generates economic profit. Informational imperfections also may do it. We may have $V(\mathcal{R}, D) < V(\mathcal{Q}, S)$ simply because no well-financed alternative proprietors are appropriately linked to the employees and customers of the firm. Indeed, innovation is simply a particular case of market imperfection, in which the market imperfection arises from "the creative response" (Schumpeter, 1947).

12.1.5 Customers

Thus far we have explored the conditions that lead to the dismissal of an employee or to a transfer of ownership, so far as these follow from the dominance relation. For completeness, something might be said about the possibility that a customer might be expelled.

Let $\iota \in B \subset S$ and consider the deviation $D = S \backslash \{\iota\}$. Here again we assume that $S \backslash (\iota)$ is feasible in that the linkages among the members of S are not disrupted by the removal of ι. It seems prima facie that the removal of one customer would reduce the consumers' surplus and so the value of the coalition. This intuition is sound but the reasoning is more complex. Recalling that

$$V^*(Q,S) = max \sum_{\kappa \in B} f_\kappa(q_\kappa) - \sum_{i \in E} g_i(h_i) \qquad (12.18)$$

subject to constraints (A11.1a–c), (A11.3d) and (A11.3e) in the appendix to the previous chapter, the parallel for D is

$$V^*(R,D) = max \sum_{\substack{\kappa \in B \\ \kappa \neq \iota}} f_\kappa(q'_\kappa) - \sum_{i \in E} g_i(h'_i) \qquad (12.19)$$

As before, we assume that the rational successor $R = \sigma(Q, D)$ is the naïve successor. Now consider

$$V^\ddagger(Q,S) = max \sum_{\kappa \in B} f_\kappa(q^*_\kappa) - \sum_{i \in E} g_i(h^*_i) \qquad (12.20)$$

subject in addition to the constraint that $q^*_\iota = 0$. By the Le Chatelier–Samuelson principle, $V^\ddagger(Q, S) \leq V^*(Q, S)$. Informally, $V^\ddagger(Q, S) = V^*(Q, S)$ only if the optimal $q^* = 0$ in 1, and this will be so only if $f_\iota(0)$ is less than the efficient marginal cost price for S. Now consider

$$V^\ddagger(R,D) = max \sum_{\substack{\kappa \in B \\ \kappa \neq \iota}} f_\kappa(q^\ddagger_\kappa) - \sum_{i \in E} g_i(h^\ddagger_i) \qquad (12.21)$$

subject, again, to the additional constraint $q^\ddagger_\iota = 0$. Since q_ι is not a variable in this problem the constraint cannot bind, and indeed identically $V^\ddagger(R, D) = V^*(R, D)$. On the other hand, the functions and constraints that define (12.20) and (12.21) are identical, so it must be that $V^\ddagger(R, D) = V^\ddagger(Q, S)$. Thus we have

$$V^*(R, D) \leq V^*(Q, S) \qquad (12.22)$$

In ordinary language, it is not as a rule profitable to drop a paying customer.

12.1.6   Interim Summary

In the context of a cooperative conception of the firm, recontracting is a model of competitive offering, and so we apply it to determine what coalitions, among the feasible and considerable coalition structures, will be stable in the face of competitive offering. Coalitions in such a structure will realize some surplus, and the surplus will be distributed according to bargaining power as described in the previous chapter. This approach gives rise to some familiar ideas from microeconomics: if an employee is paid more than her or his marginal productivity, the coalition structure will be dominated by one in which she or he is separated from the coalition; customers, however, will not ordinarily be separated; under strong competition among proprietors, economic profit is reduced to zero; but informational imperfections (including those created by entrepreneurship) can stabilize positive profits. These results do not require us to assume that labor or capital are homogenous inputs. All of this discussion so far assumes information sharing links, and thus the set of feasible coalition structures, are given.

12.2   SEARCH AND ESTABLISHMENT OF LINKS

However, the links are not given for all time, but established through processes of search. Search has a cost – if not, all possible links would be formed, and we would be living in a world of universal "perfect competition" (or perhaps in a world without any stable equilibria, if the core of the market game is empty, as it is in some models.) Further, the decision to assume the cost of search can only be made noncooperatively, since, *pari passu*, links required for cooperative decisions do not yet exist. However, it is not quite so simple. Individual "entrepreneurs" may search for new employees and customers to form a new coalition. In such an entrepreneurial framework, all decisions are individual and so purely noncooperative in this sense. But much of the search for new employees and customers (and capital partners) is carried on by ongoing coalitions. In a fundamentally cooperative perspective, the decision to commit resources to search will be made cooperatively for the common benefit of the existing members of the coalition – the *insiders*.[3]

In the next chapter, this discussion will provide the basis for a reconsideration of the implications of monopoly and monopsony power in the context of public policy. Here we will parallel the very influential Nobel laureate work of Pissarides and Mortensen (see, for example, Mortensen and Pissarides, 1994) and many others in the different context of this chapter.

12.2.1   Search in Labor Markets

First, consider the "labor markets," that is, the formation of new links to potential employees. The mutual benefit of such a new link will depend on the idiosyncratic productivity of the new employee in the event that she or he does join the coalition. In the absence of an information-sharing link, this parameter is unknown, and a major reason motivating search, on both sides, is to find matches in which the idiosyncratic productivity is relatively high.

Let the individual worker be i and the coalition j and the idiosyncratic productivity of i in j be $k_{i,j}$. We may suppose that, *ex ante*, $k_{i,j}$ is drawn from a joint probability distribution $\phi(i,j)$ and that the marginal distribution $\phi(j|i)$ is known to individual i and the marginal distribution $\phi(i|j)$ is known to the potential employer coalition j.

Suppose then, that the coalition j calculates that, as a result of recruitment, its surplus will be increased in the common interest of the members. Following the large literature on search and matching in labor markets, we assume that a link will be formed (if at all) with a random draw from those workers who are seeking work. Similarly, suppose that individual i is seeking work. By entering the "labor market" individual i may expect, if successful, to form a link with a coalition drawn at random from all those coalitions seeking new employees.

We might model these processes in an "as if" sense along the following lines. There is a single employment exchange. In order to seek employment, an individual must register with the employment exchange, at some cost. A coalition seeking new members sends a message to the employment exchange, which redirects the message to one of the individuals seeking employment at random, with equal probability for each registered worker. Let J be the number of messages sent and U the number registered as seeking work.

Suppose then that a worker forms a link only with the coalition that has sent the first message that the worker receives, while any other messages go without any response. To a coalition considering sending a J + first marginal message, the probability that a particular worker i has not yet received a message is

$$p = \left(1 - \frac{1}{U}\right)^{J} \tag{12.23}$$

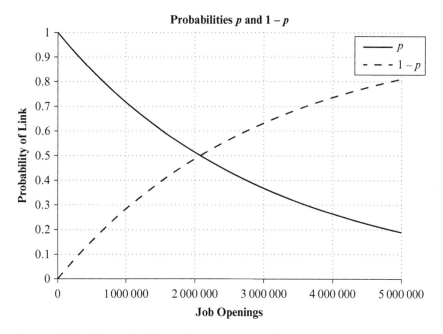

Figure 12.2   The probability of success in search

and accordingly this is the proportion who are still available and the probability that the marginal message results in a new link. Since U and J are quite large numbers, an example will be helpful to suggest a reasonable scale for p. This is shown in Figure 12.2, which assumes three million workers are available and shows the variation in p as the number of job openings varies from zero to five million. Figure 12.3 shows the values of p – we might say, of the slackness of the labor market[4] – for the United States, for the current century, treating the level of unemployment as U and the level of job openings as J, and using data from the Bureau of Labor Statistics, including the JOLTS project. If nothing else, p seems reasonable as an index of the slackness of the US labor market in the current century. The tighter the labor markets, then, the greater is the probability that an additional search will be a wasted cost.

From the point of view of the potential employee, the probability of receiving at least one message and thus forming a link is precisely $1 - p$. Thus, as labor markets grow tighter, the probability of a successful search for the potential employee increases.

If a link is formed, then an employment relation *may or may not* occur. The new coalition created by the employment relation may be dominated

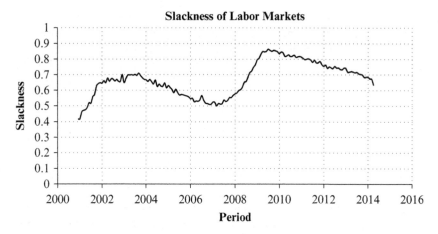

Figure 12.3   Values of p from Bureau of Labor Statistics data

either by the unexpanded coalition or by the singleton coalition $\{i\}$. Let V be the value created by the unexpanded coalition and V^+ by the union of the unexpanded coalition with $\{i\}$. Then V^+ will depend on $k_{i,j}$. If $V^+ < V$, then there is no mutual benefit from the expansion of the coalition – worker i will not be hired. If $V^+ > V$, then it is possible for both the insiders and the new employee to be better off than they would be if the hire did not occur. The payoffs to insiders and to the new hire i will depend on the distribution of bargaining power in the new coalition. Some insiders may have less bargaining power in the new coalition, but, as argued in subsection 12.1.2, bargaining power will not shift in such a way that an insider would be worse off, as that would prevent the dominant coalition from forming. Nevertheless, $\beta_{i,j}$ will be undetermined within some range bounded by zero and $V^+ - V$. In other words, $\beta_{i,j}$, like $k_{i,j}$, is idiosyncratic to the match between i and j.

For worker i, let γ_i be the cost of registering for the employment exchange. The cost γ_i may vary from one worker to another; for example, it may be more costly for a person who is currently employed to register than for one who is not. The worker will use his knowledge of $\phi(j|i)$ and whatever information she or he may have about the tightness of the labor market and the variation of idiosyncratic bargaining power to determine an expected value of y_i conditional on entering the labor market. If $y_i > \gamma_i$ then she or he will enter the labor market. If $\gamma_i > y_i > 0$, then i will be a "discouraged worker." If $y_i < 0$, then the reasonable choice for i is to leave the labor force. On this basis potential employees will enter, depressing $\frac{J}{U}$, until the expected value $[1 - (1 - \frac{1}{U})^J]y_i$ is no more than γ_i.

Potential employer coalitions (including entrepreneur singletons) will form their respective expected values of V^+ depending on the distributions $\phi(i|j)$ and the other information available to them, but also on $\frac{J}{U}$. Potential employers will continue to enter, raising $\frac{J}{U}$, until the expected value $(1 - \frac{1}{U})^J (V^+ - V)$ is no more than zero. These conditions will be approximated by the large-group equilibria discussed in much of the literature on search and matching in labor markets.

The condition for equilibrium in the entry of potential employers into labor markets deserves further comment. The condition is that $(1 - \frac{1}{U})^J (V^+ - V)$ be zero; but $(1 - \frac{1}{U})^J$ cannot be zero for finite J. However, V^+ will not be independent of $(1 - \frac{1}{U})^J$. The cost of recruiting is an offset against the surplus generated by the expanded coalition. Thus, as J increases, V^+ declines, and for a large enough J (given U), $V^+ - V$ may be zero or negative. This will be further considered in the next chapter.

Of course, this is a very simplified model of labor market search. However, in common with much of the literature, it captures what seem to be important features of real labor markets. First, matching of workers to jobs is not free, and that is why some, on both sides, remain unmatched. Second, a marginal search by an employer, by increasing labor market tightness by some very small degree, generates two kinds of externalities: one, to other employers, a negative externality in reducing the productivity of their search; and, second, to workers, a positive externality in increasing the productivity of their search. Similarly, the entry of a marginal worker, by reducing the tightness of labor markets, similarly generates two opposite externalities. In general these externalities will not be symmetrical, but depending on the details may justify policies such as active labor market policies or subsidies to the unemployed, as promoting efficiency. There is a large and much honored literature on this topic, and further formal discussion of it will be beyond the scope of this part of the book.

Most of that literature, however, assumes noncooperative decision-making and, in particular, profit maximization. For the model in the previous chapter, profit maximization means that the bargaining power of any employee, $\beta_{i,j}$, is zero. It follows that $y_i = \beta_{i,j} V^+ = 0$ and the expected benefit from entering the labor market is $-\gamma_i$ and a rational person will not enter the labor market. (This is an instance of the no-action equilibrium illustrated at McCain, 2013, p. 145.) In a world of costly search and profit maximization, there will be no labor markets.

What this means is that successful firms will find ways to commit themselves to routines and rules that limit the bargaining power of insiders in general, and the proprietor in particular, so long as it is in the interest of the insiders to expand. If it is not in the interest of insiders to expand, the firm will not be recruiting in labor markets.

12.2.2   Search for New Customers

As Hall (2008) observes, customers, no less than employees, arrive at an exchange relation with a firm through a process of costly search and matching. In this case, activities undertaken by firms to create links include advertising and other forms of marketing, while individuals' search for new customer relationships is called "shopping." Following *very roughly* Hall's procedure, we would apply to customer search a model similar to that applied to labor markets, characterizing market tightness in a similar way. However, there are some important differences between consumer links and employee links, in this model. In particular, a customer may be linked to more than one coalition for production and exchange, and probably to many. Thus it is not plausible to suppose that a potential consumer responds only to the first message. It is also important that a potential consumer can make more contacts by allocating more of her or his resources (time) to shopping. Accordingly, the information-transfer metaphor will be a little more complex in this case.

Once again we envision a central messaging system to which a coalition or entrepreneur may send messages, and the messages are redistributed at random. However, instead of redistribution to individual customers, a message is sent at random to one of Λ message boxes. In turn, these boxes with the messages they contain may be claimed by a potential customer. However, (1) a potential customer may claim more than one message box, and (2) since a customer may be linked to more than one seller, all messages in any box claimed by consumer ι will result in a link between the seller and ι. The number of boxes claimed by ι will be proportionate to the resources (time) that ι devotes to shopping. Suppose that a representative consumer ι claims v_ι boxes and firm j sends v_j messages. The v_j messages are distributed at random without replacement among the boxes. Nevertheless, more than one of the v_ι boxes claimed by ι may contain messages from i.

A message sent by firm j will form a new link if (1) it is sent to a box that is claimed by some potential customer, and (2) it does not duplicate a message in another box claimed by the same consumer. (This assumes the recipient is not already a member of coalition j. We will ignore this consideration for simplicity; perhaps we may reason that current members are a negligibly small proportion of the population of potential customers.) If there are Φ representative consumers ι each of which claims v_ι boxes, then $v_\iota \Phi < \Lambda$ boxes are claimed. (No box is claimed by more than one consumer.) Thus, the probability that a single message reaches a claimed box is $\frac{v_i \Phi}{\Lambda}$ and the probability that it is nonduplicative is $(1 - \frac{v_i}{\Lambda})^{v_j-1}$, where v_j is the number of messages sent by firm j. The product of these,

$$p = \frac{v_i \Phi}{\Lambda} \left(1 - \frac{v_i}{\Lambda}\right)^{v_j-1} \qquad (12.24)$$

is the probability that the v_jth message will form a new link. Suppose that firm j sends v_j messages to a sample of boxes chosen without replacement. Then the marginal productivity of messages sent is p times the expected incremental surplus as a result of the establishment of the link. Since $(1 - \frac{v_i}{\Lambda}) < 1$, p will decrease with increasing v_j. Consider Figure 12.4. The solid line shows the values of p for $v_i = 4$, $\Lambda = 50,000$ and $\Phi = 10,000$. The dashed line shows the values of p for $\Phi = 5000$ (that is, half as many potential customers). In Figure 12.5 the same are shown with $\Lambda = 100,000$.

Once again, this is a very simplified information process, but it probably captures some important aspects of the search and matching of sellers and buyers in the real world. First, the more resources devoted to search by each side, the more links will be formed, subject to diminishing returns. Second, some advertising goes to waste because it is redundant. The dimension of the information space Λ is an arbitrary parameter

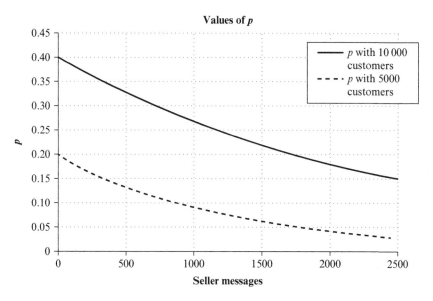

Figure 12.4   The probability that the next message will attract a new customer

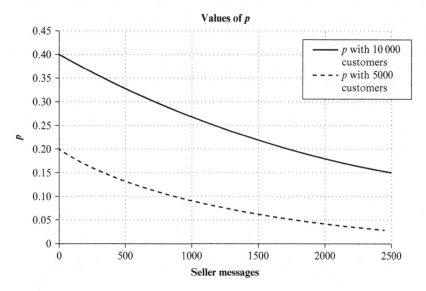

Figure 12.5   The same with $\Lambda = 100,000$

representing a completely unstructured process of message distribution. In real markets there probably is a great deal of structure. Most marketing is at least somewhat targeted. Targeting would seem likely to increase the productivity of marketing expenditure (as would a decrease of Λ in this model) but also to increase the probability of redundant messages (as would an increase of v_t in this model). Thus, a modestly more complex model might be useful for some purposes. In recent history, the invention of e-mail decreased the cost per message, but seems also to have increased the redundancy of advertising messages (to judge from my junk mail file, anyway). But such a more realistic model will be beyond the scope of the book.

12.3   CONCLUSION

The array of coalitions actually formed in an economy as conceived here will reflect two considerations, one noncooperative and one cooperative. The noncooperative influence is the search and matching process by which new informational links are formed. This is unavoidably noncooperative, since cooperative decisions cannot be made until the links exist. As a result, it may allow for inefficient phenomena such as involuntary unemployment, unfilled positions, and externalities. The cooperative consideration is a process of recontracting or competitive offering among those who, being linked, are potentially members of common coalitions. This process gives rise to some of the conditions familiar from neoclassical economics, such as the value of marginal product as the upper limit of an individual employee's wage and the zero profit condition in full competitive equilibrium.

NOTES

1. The definition of the rational successor function and some other formal details including proof of existence and uniqueness are given in the appendix to this chapter.
2. See the appendix to this chapter for details.
3. There is a large literature on the implications of insider interests in macroeconomics. See, for example, Lindbeck and Snower (2001). Their approach, however, is somewhat ambiguous as to whether decisions are made cooperatively or noncooperatively.
4. This interpretation differs from that in the received literature which treated J/U as a measure of the slackness of labor markets, but, as we will see in Chapter 14, the two indices agree qualitatively.

APPENDIX A12

For this chapter, we consider a coalition structure to be feasible if (1) it comprises coalitions that are feasible, in the sense that their members are appropriately linked and (2) none are incompatible. In addition, formally, it will be helpful if every feasible coalition structure includes every possible singleton, as well as the null set, and that is assumed. Otherwise, an agent who happens to be unlinked would not be a member of any coalition. Accordingly, let $\mathcal{M}$ be the set of all feasible coalitions, and $|S| < 2 \Rightarrow S \in \mathcal{Q}$. Let $\mathcal{N}$ be the set of all unordered pairs of coalitions that are incompatible; $\mathcal{N} \subset \{(S_1,S_2)|S_1 \subseteq N, S_2 \subseteq N\}$ (where N is the set of agents in the game). Further, denote $\mathcal{N}_S = \{C|(C,S) \in \mathcal{N}\}$, where $S \subseteq N$, $C \subseteq N$. Then, denoting $\mathcal{Q}$ as a coalition structure (that is, an arbitrary set of subsets of N) and S as the set of all feasible coalition structures, $\mathcal{Q} \in$ S if

(1) $S \in \mathcal{Q} \Rightarrow S \in \mathcal{M}$
(2) $S_1 \in \mathcal{Q}, S_2 \in \mathcal{Q} \Rightarrow (S_1,S_2) \notin \mathcal{N}$
(3) $|S| < 2 \Rightarrow S \in \mathcal{Q}$

Note that when feasibility is defined in this way, partitions in general may not be feasible, since an arbitrary partition may or may not be feasible. Thus, what follows will differ from McCain (2013, pp. 50–51), despite some similarity of terminology.

An FCS game Γ comprises a set of feasible coalitions S together with a value function for coalitions. Since the value of S will depend on the coalition function $\mathcal{Q}$ in which it is imbedded, the value created by $S \in \mathcal{Q}$ will be denoted as $V(S, \mathcal{Q})$. Drawing on the previous chapter $V(S, \mathcal{Q})$ is assumed to be a constant regardless of the distribution of $V(S, \mathcal{Q})$ among the members of S. Thus, an imputation for $\mathcal{Q}$ is *admissible* if $\forall S \in \mathcal{Q}, \Sigma_{i \in C} y_i = V(S,\mathcal{Q})$. A candidate solution is a pair comprising $\mathcal{Q} \in$ S and $\mathbf{y} = \{y_i\}_{i \in N}$ admissible for $\mathcal{Q}$. Now suppose $\exists \mathcal{R} \in \mathcal{S}$ and $C \in \mathcal{R}, C \notin \mathcal{Q}$, and $V(C,\mathcal{R}) > \Sigma_{i \in N} y_i$. Then $(\mathcal{Q},\mathbf{y})$ is *dominated by* $\mathcal{R}$ *via* C. Suppose further that, for any $\mathbf{y}$ admissible for $\mathcal{Q}, \exists C \in \mathcal{R} \ni (\mathcal{Q}, \mathbf{y})$ is dominated by $\mathcal{R}$ via C. Then $\mathcal{Q}$ is unconditionally dominated (u.d.) by $\mathcal{R}$. For example, suppose that the value function is strictly superadditive and $C = C_1 \cup C_2$, where $C_1 \in \mathcal{Q}, C_2 \in \mathcal{Q}$. Then for any $\mathbf{y}$ admissible for $\mathcal{Q}, V(C, \mathcal{R}) > V(C_1, \mathcal{Q}) + V(C_2, \mathcal{Q}) = \Sigma_{i \in C} y_i$, so that $\mathcal{R}$ u.d. $\mathcal{Q}$.

However, the fact that $\mathcal{R}$ dominates or u.d. $\mathcal{Q}$ does not necessarily mean that $\mathcal{Q}$ will give way to $\mathcal{R}$ if C is formed, since $\mathcal{R}$ may itself give rise to some further reorganization or may require some reorganization of $N \backslash C$ that is unlikely to occur. Before participating in a deviation such as C, a rational agent would attempt to anticipate these reorganizations. This anticipation could be expressed in terms of a *successor function* (McCain, 2013, pp. 49–51) $\mathcal{R} = \sigma\,(Q,C)$. If C is formed and no other unavoidable reorganizations occur, $\sigma^*(Q,C) = \{C\}\cup Q\backslash\mathcal{N}_C$. Then σ^* is the *naïve successor* of Q via C.

Then define $\mathcal{R} \in$ S is a *rational successor* of Q via C if

(1) $C \in \mathcal{R}$
(2) $\exists$ a sequence $\mathcal{P}_1, \mathcal{P}_2, \ldots, \mathcal{P}_r \ni$
   a. $\forall i = 1,\ldots,\text{r},\ \mathcal{P}_i \in \mathcal{G},\ \text{S} \in \mathcal{P}_i$
   b. $\mathcal{P}_1 = \sigma^*(Q,S)$
   c. $\mathcal{P}_r = \mathcal{R}$
   d. $\forall i < r,\ \mathcal{P}_{i+1}$ u.d. $\mathcal{P}_i$
   e. $\forall i = 1, \ldots ,\text{r}$, if $\exists\, \mathcal{U} \in \mathcal{G} \ni \text{S} \in \mathcal{U},\ \mathcal{U}$ u.d. $\mathcal{P}_i,\ \mathcal{U} \neq \mathcal{P}_{i+1}$ then $\exists$ a sequence $\mathcal{W}_1, \mathcal{W}_2, \ldots, \mathcal{W}_s, \ni \forall \text{j} = 2,\ldots,\text{s},\ \mathcal{W}_j$ u.d. $\mathcal{W}_{j-1}$, $\mathcal{W}_1$ u.d. $\mathcal{U}$, and $\mathcal{U}$ u.d. $\mathcal{W}_s$.
   f. Conversely, if $\mathcal{U} \in \mathcal{G} \ni \text{S} \in \mathcal{U}$, and $\exists$ a sequence such as in e), then $\mathcal{U} \notin \{\mathcal{P}_1, \ldots \mathcal{P}_r\}$

Remark: The conditions say that (1) we are concerned with reorganizations that occur when S is given; (2a–d) that $\mathcal{R}$ is a product of a sequence of reorganizations of $\sigma^*(\,Q,S)$, each of which makes some group better off than they can be in the previous organization; (2e–f) that a reorganization will *not* take place if it creates ambiguity in either of two ways: (1) that the reorganization is part of a dominance cycle or (2) that it is u.d. by two or more distinct further reorganizations, so that follow-on reorganizations are unpredictable. These conditions might be thought of as expressing a kind of aversion to uncertainty. In this case the uncertainty cannot be reduced to risk without at least some further assumptions. Thus these may be thought of as instances of "Knightian uncertainty" (Knight, 1921). It is not asserted that an irrational successor could never occur, but that a rational agent, knowing also the rationality of other agents, would not anticipate its occurrence.

Theorem: For any FCS game Γ, $S \in \mathcal{M}$, $S \notin Q$, $Q \in$ S, a rational successor $\sigma(Q,S)$ exists and is unique.

Existence. If a sequence as in (2a) exists with $r > 1$, then $\mathcal{R} = \sigma(Q,S) = \mathcal{P}_r$; otherwise, $r = 1$ and $\mathcal{R} = \sigma(Q,S) = \sigma^*(Q,S)$.

Uniqueness: Suppose $\exists\ \mathcal{R}^1 \neq \mathcal{R}^2$ satisfying (1) and (2) above, with $\mathcal{P}_j^1$ the sequence corresponding to $\mathcal{R}^1$ and $\mathcal{P}_j^2$ the sequence corresponding to $\mathcal{R}^2$.

Lemma: $\exists\ \mathcal{P}_j^*$ a common element of both sequences. *Proof*: by (2b), $\sigma^*(Q,S)$ is a common element.

Let $\mathcal{P}_j^*$ be the common element with the largest index.

Lemma: $\exists\mathcal{P}_{j+1}^1$. *Proof*: Suppose not. Then $\mathcal{P}_j^* = \mathcal{R}^1$. Then since $\mathcal{R}^1 \neq \mathcal{R}^2$, $\exists\mathcal{P}_{j+1}^2$ with $\mathcal{P}_{j+1}^2$ u.d. $\mathcal{P}_j^*$. By (2e) $\mathcal{P}_{j+1}^2$ is part of a dominance cycle, but by (2f) this means that $\mathcal{P}_{j+1}^2 \notin \{\mathcal{P}_i^2, \dots \mathcal{P}_r^2\}$, a contradiction. A similar argument establishes that $\exists\mathcal{P}_{j+1}^2$.

Lemma: $\mathcal{P}_{j+1}^1 = \mathcal{P}_{j+1}^2$. *Proof*. If not, then by (2e) one or the other of $\mathcal{P}_{j+1}^1$, $\mathcal{P}_{j+1}^2$ must be an element of a dominance cycle, and therefore, by (2f), is not part of the corresponding sequence, contradiction.

It follows however that $\mathcal{P}_{j+1}^1 = \mathcal{P}_{j+1}^2$ is a common element of both sequences, contradicting the assumption that j is the largest element of a common sequence.

Remark: what this argument says is that for a sequence leading to a rational successor, each reorganization in the sequence is unique, so in particular this applies to the last step, the rational successor.

Given Q, S as before, the rational successor will be denoted as $\sigma(Q, S)$. Now, let $(Q, \mathbf{y})$ be a candidate solution and suppose $\exists S \in \mathcal{M} \ni (Q, \mathbf{y})$ is dominated by $\sigma(Q, S)$ via S. Then $(Q, \mathbf{y})$ is *unstable*; otherwise it is stable. Then the core of FCS Γ comprises those candidate solutions in Γ that are stable.

13. Monopoly and monopsony revisited

The previous two chapters have explored several aspects of an economic model that treats the business firm as a cooperative coalition with membership limited to agents who are linked by information-sharing links. Chapter 11 addresses the idea that rational cooperators will make decisions that are Pareto-optimal as among themselves. One implication is that, even though the coalition is both a monopolist in respect of its customer-members and a monopsonist in respect to its employees, the traditional analysis of monopoly and monopsony will not apply. This traditional analysis implies that the allocation of resources among employees, customers and proprietors will be inefficient, and this inefficiency, in itself, means that the analysis cannot apply to a group of rational cooperators comprising all those who jointly benefit from the activities of the firm. Chapter 12 considered two kinds of decisions that would further determine the coalition structure of an economy conceived in this way. In particular, the second section of Chapter 12 considers the *noncooperative* process of search and matching by which new links are made that may lead to the expansion of the coalition. The decisions to enter into this matching process are unavoidably noncooperative, since they are made before the informational links are made that might support cooperative decisions. Nevertheless, the decisions to seek new employee-members of the coalition, and those to seek new customer-members of the coalition, will be made cooperatively by the ongoing members of the coalition. In labor markets, it is often observed that there may be differences of interest between *insiders* and *outsiders*; that is no less true of insider and outsider consumers. Accordingly, this chapter returns to the cooperative decisions of the insiders of the coalition, allowing in addition for the decisions to attempt to recruit new members.

At the same time we reconsider the implications of monopoly and monopsony power. A major objective of this discussion is to permit a direct comparison of the implications of cooperative firms with the traditional analysis in the case of monopoly and monopsony power, and the simplifying assumptions made are chosen accordingly. We begin with recruitment of new employees and thus with monopsony power. In each case, we first consider an intermediate case with costs of recruitment but without market power.

13.1 MONOPSONY POWER

We now consider the decision of a coalition to recruit new employee members. This is, of course, an inherently uncertain prospect. The parameters of a newly linked potential employee, k_i and β_i, are not known until the link with the individual is formed, and this uncertainty is the motive for the costly search that is part of the recruiting process. For simplicity, and for comparability with the traditional "theory of the firm," this decision will be treated as a certainty equivalent decision to recruit representative new employees. Suppose, then, that a representative new recruit will have productivity parameter k, and accordingly will be assigned h units of effort, and, with the bargaining power that can be anticipated, will be paid a (wage) side payment z. The decision will reflect the interests of "insider" continuing employees as well as continuing members in other categories; but in any case the payment z to the representative new employee reduces the surplus available to the insiders as a group.

Now, suppose that, by diverting L_1 units of labor from production to recruiting, the coalition can recruit an expected value of m of new employees. (As an expected value m need not be a whole number.) This increases the expected effective labor power of the coalition by $\ell =$ mhk units at the cost of increasing the labor overhead from L_0 to $L_0 + L_1$. Assuming that the relation of L_1 and l is one-to-one, denote L_1 as r(ℓ). For $\ell \leq 0$ we will have r(ℓ) = 0; for $\ell > 0$, it will be assumed that $r(\ell) > 0, \frac{\partial r}{\partial \ell} > 0, \frac{\partial^2 r}{\partial \ell^2} > 0$. In general r($\ell$) will depend on the conditions of the labor market: tightness and the representative parameters k and z, and on the conditions of the recruiting coalition. Accordingly, r will shift from period to period and will differ from firm to firm.

13.1.1 Intermediate Case

To represent the decision of a coalition to commit r(ℓ) of their resources to recruitment to add an expected value of ℓ to their effective labor, we revisit the optimum model in the appendix to Chapter 11. For this purpose the disagreement payoffs w_i will be the payoffs expected in the case of no recruitment, not the insiders' outside options, since in this case "no decision" means no recruiting. We will first consider an intermediate case in which there are recruitment costs but no monopsony power. In the appendix to Chapter 11, the labor resource allocated to production is

$$L \leq \sum_{i \in E} k_i h_i - L_0 \tag{13.1}$$

where L_0 is a labor overhead comprising supervisory labor and other labor allocated to purposes other than the production of goods and services to satisfy the wants of customers. Within this category are activities of recruiting new employees ("human resources") and recruiting new customers ("marketing"). The decision to recruit is made at an earlier (noncooperative) stage of the multistage decision process than is the decision on the allocation of that labor, but the succession of stages is logical rather than chronological and describes an ongoing process. In any case a larger workforce in production will require a larger allocation of labor to recruit and maintain that workforce as links spontaneously disappear through deaths, retirements, and quits. We may suppose that, in an ongoing firm, the labor set aside for recruiting is insider labor. In any case, the insiders of the coalition will decide how much of its labor power to allocate to recruiting in such a way as to increase their own surplus.

Allowing for recruitment, in place of (13.1) we have

$$L \leq \sum_{i \in E} k_i h_i + \ell - \imath(\ell) - L_0 \tag{13.2}$$

This is equation (A13.1a) in the appendix to this chapter. Again applying the bargaining model to determine the insiders' decision to recruit, for this intermediate case, if $\ell > 0$, that is, if the coalition does recruit, it will allocate resources to recruiting up to the point that

$$\frac{\partial f_1}{\partial q_1} \frac{\partial F}{\partial L} \left(1 - \frac{\partial \imath}{\partial \ell} \right) = \frac{z}{hk} = Z \tag{13.3}$$

This is equation (A13.2d) in the appendix. On its face this is a straightforward extension of the traditional theory, since $Z = \frac{z}{hk}$ is the wage cost of one unit of effective labor from a representative new recruit, and $\frac{\partial f_1}{\partial q_1} \frac{\partial F}{\partial L}$ is the value of the marginal product of effective labor, evaluated at the common marginal utility of the firm's output. In a perfectly competitive model $\frac{\partial \imath}{\partial \ell} = 0$, so 3 would say that recruitment is carried on to the point that that the VMP is equal to the wage cost per unit of effective labor. More generally, we replace the VMP with the VMP *net of the labor cost of recruiting a marginal unit of effective labor*.

Recall, however, that this is a condition on *newly recruited* labor. That is, it characterizes the rate of recruitment of new labor, not the allocation of the labor resource, as it would in a neoclassical model. Moreover, in given circumstances the coalition may not recruit, but nevertheless not dismiss insider employees – it may reduce its labor force only by attrition, or not at all. This corresponds to a corner solution, as expression (A13.2e), if in addition (drawing on Chapter 12, subsection 12.1.3) the insider employees are paid no more than their marginal value product.

13.1.2 The Case of Positive Monopsony Power

In a neoclassical economic model, monopsony power results in misallocation of resources in that the profit-maximizing employer restricts employment to something less than an efficient rate (Robinson, 1933). This occurs because (by assumption) a "law of one price" is associated with labor markets in a neoclassical model. In the presence of wage discrimination – which is assumed in the cooperative model – monopsony power will have no necessary impact on resource allocation. However, wage discrimination (like cooperation) requires information that will not be available until the link between the potential employer and the potential employee is made. If the relevant labor market is limited by comparison with the scale of the coalition's recruiting, the parameters corresponding to a representative new employee, k and z, may vary with ℓ, as also the tightness of the labor market may. For present purposes, and to parallel traditional labor market theory as nearly as possible, suppose only that z rises with ℓ. That is, $z = h(\ell)$ with $\frac{\partial h}{\partial \ell} > 0$. The necessary condition for an optimum positive rate of recruitment of new employees will then be

$$\frac{\partial f_1}{\partial q_1} \frac{\partial F}{\partial L} \left(1 - \frac{\partial z}{\partial \ell} \right) = \frac{z}{hk} + \frac{\ell}{hk} \frac{\partial h}{\partial \ell} \tag{13.4}$$

This is equation (A13.3c) in the appendix to this chapter. Once again, if $\frac{\partial z}{\partial \ell} = 0$, we have just the traditional result – recruitment is carried only to the point that the VMP is equal to the marginal resource cost (MRC), that is, the increase in the cost of (newly recruited) labor power with the recruitment of one additional unit, which will be greater than the wage cost of the one additional unit. Allowing for $\frac{\partial z}{\partial \ell} > 0$, we say instead that the VMP *net of* the labor required to recruit an additional unit of labor is equal to the MRC.

Note that

$$\ell \approx 0 \Rightarrow \frac{z}{hk} + \frac{\ell}{hk} \frac{\partial h}{\partial \ell} \approx \frac{z}{hk} \tag{13.5}$$

so that a corner solution will again correspond to the condition (A13.2e).

This example is to be distinguished from the traditional one in several important ways, however. First, it applies to the rate of recruitment of new labor, not to the use of labor overall. Second, in this more complex model, there are several different ways in which monopsony power might lead to modifications. Like the parameter z, the productivity parameter k might change – perhaps decline – as ℓ rises. Further, in a smaller labor market relative to ℓ, the marginal cost of recruitment, $\frac{\partial z}{\partial \ell}$, would probably rise more rapidly than it would in a larger labor market.

13.1.3 Minimum Wage Laws

The most important policy application of monopsony theory is to minimum wage regulations. Proposals for minimum wage legislation are often rationalized by the supposition that employers have monopsony power (Robinson, 1933). In the traditional theory, a moderate lower limit on the permissible wage can increase the firm's employment (by reducing the effective $\frac{\partial \ell}{\partial \ell}$ to zero). In the analysis of this chapter the "law of one price" in labor markets is not a result of informed competition but of averaging a sample of the set of workers seeking jobs. Thus $h(\ell)$ might be an average over workers some of whom expect wages at the regulated minimum and some above it. If so, however, it seems likely that wage regulations might decrease the absolute value of $\frac{\partial \ell}{\partial \ell}$, if not to zero, for any given ℓ (while also raising z). This might offset monopsony power and increase recruiting somewhat.

However, minimum wage regulation will apply to insiders no less than new recruits. Thus, if the minimum wage permitted is z_0, we could model the effect of the minimum wage on insiders by adding to (A11.4b) or (A13.2a) a set of constraints $z_i \geq z_0$, $\forall i \in E$. In the Lagrangean function we will have an additional term

$$\sum_{i \in E} \mu_{i,6}(z_i - z_0) \tag{13.6a}$$

Relative to (A11.6a) this will yield

$$\forall i \in E, \frac{\partial \mathcal{L}}{\partial z_i} = \frac{\beta_i}{y_i} - \mu_4 + \mu_{i,6} \tag{13.6b}$$

Now, constraint (13.6) may or may not be binding for a particular employee. For an employee with a relatively great individual bargaining power β_i or a relatively good outside option, we may have

$$z_i > z_0 \tag{13.6c}$$

In that case the necessary condition

$$\mu_{i,6}(z_i - z_0) \leq 0 \tag{13.6d}$$

requires $\mu_{i,6} = 0$. When the constraint is binding, that is, $z_i = z_0$, we may have $\mu_{i,6} > 0$.

Letting E_1 be the set of employees for which the constraint is binding and E_2 the set for which it is not, we have in place of (A11.6d)

$$\forall i \in N \backslash E_1, \beta_i = \mu_4 y_i \tag{13.7a}$$

$$\forall i \in E_1, \beta_i = (\mu_4 - \mu_{i,6}) y_i; \tag{13.7b}$$

in place of (A11.8e)

$$\mu_4 = \frac{\sum_{i \in N} \beta_i + \sum_{i \in E_1} \mu_{i,6} y_i}{V} \tag{13.7c}$$

So that, in place of (A11.6d), we have

$$\forall i \in N \backslash E_1, y_i = \frac{\beta_i}{\mu_4} \tag{13.7d}$$

$$\forall i \in E_1, y_i = \frac{\beta_i}{\mu_4 - \mu_{i,6}} \tag{13.7e}$$

We see that the effective relative bargaining power of those who are constrained is *increased*, relative to their intrinsic bargaining power β_i. In short, the principal effect of a minimum wage law on the insiders of the coalition is to increase the proportion of the surplus to those who would otherwise be paid less than the minimum. It will compress the wage distribution, raising the lowest but probably also reducing the highest, along with some reduction of economic profit and consumers' surpluses.

13.1.4 Entry Equilibrium in Labor Markets

We recall from Chapter 12.2 that a condition for a noncooperative equilibrium in the decision of a coalition to increase recruitment of additional labor is that the expected value of an additional recruit be zero, that is

$$\left(1 - \frac{1}{U}\right)^J (V^+ - V) = 0 \tag{13.8}$$

this in turn requires that

$$(V^+ - V) = 0 \tag{13.9}$$

We now consider how the net benefit of hiring will vary with the rate of recruitment. The cost of recruiting, $r(\ell)$, will also depend on the tightness of labor markets. Thus, in place of (A13.1a) in the appendix, write

$$L \le \sum_{i \in E} k_i h_i + \ell - r\left(\ell,\left(1 - \frac{1}{U}\right)^J\right) - L_0 \tag{13.10}$$

where E is the set of ongoing insider employees, ℓ the expected value of the increase in overall labor power, $r(\ell,(1 - \frac{1}{U})^J)$ the increase in overhead labor necessary to yield 1 on the average, and r is an increasing function of both variables. As J increases, $r(\ell,(1 - \frac{1}{U})^J)$ shifts upward relative to ℓ, increasing $r(\ell,(1 - \frac{1}{U})^J)$ for any recruiting target ℓ as markets become tighter; as U increases $r(\ell,(1 - \frac{1}{U})^J)$ shifts downward.

Returning to expression (13.9), as labor markets grow more tight with increasing J then for given recruiting cost ℓ declines (the constraint binds in circumstances in which it otherwise would not) so that V^+ decreases. This leads to the equilibrium condition that $V^+ \le V$ for the marginal firm entering the labor market.

13.2   MONOPOLY POWER

We now consider the decision of the coalition to recruit more customer members. Once again the parameters anticipated for a new customer are uncertain and will be considered as certainty equivalent expected values for a representative new customer. We suppose that, in an anticipated new optimum, the representative new customer would take q of the firm's output and make a side payment z into the revenues of the coalition. Suppose then that the commitment of r units of incremental labor overhead will on the average bring b new customers. This will reduce the output available to insider customers by bq (in addition to whatever reduction results from diverting r of labor from production, and except insofar as increased effort may partially offset these deductions) and will increment the revenues of the coalition by bz.

13.2.1   Intermediate Case

Once again we first consider an intermediate case with variable recruitment costs but without monopoly power. Thus, in place of (A11.1a) we have

$$L \le \sum_{i \in E} k_i h_i - r(\ell) - L_0 \tag{13.11}$$

Again applying the bargaining model to the decision of the insiders to recruit new customers, we find that the necessary condition for an optimum positive rate of recruitment of new customers then is

$$z = \frac{\partial g_i}{\partial h_i}\frac{\partial \pi}{\partial \ell} + \frac{\partial f_1}{\partial q_1} q \tag{13.12}$$

(using A11.7h and A11.7j, which are unchanged). This is equation (A13.6b) in the appendix to this chapter. That is, the recruitment of new customers is carried on until the additional revenue gained per new customer is no more than the marginal labor cost of recruiting customers, evaluated at the common marginal disutility of labor, plus the output diverted from insider customers, at the margin, evaluated at the common marginal utility of the firm's output. If we understand the right-hand side as the marginal cost of serving one additional customer, then this seems a natural extension of the neoclassical approach. Once again, however, we may not assume an interior solution. We may have a corner solution with no attempt to attract new customers. This may be uncommon, but see Lewis (1942) for a possible instance.

13.2.2 Positive Monopoly Power

In parallel to the previous discussion, we interpret monopoly power to mean that the firm faces a limited market and can increase its rate of recruiting only by means that lead to a subsample of the population less favorable for the coalition's insiders. For example, if the firm's product is differentiated by location, increasing the rate of recruiting might require drawing new customers from an increased relative distance, so that the expected average z would decline. Accordingly, represent monopoly power by

$$z = \ell(\ell) \text{ with } \frac{\partial \ell}{\partial \ell} < 0 \tag{13.13a}$$

Then, in place of (13.5), assuming an interior solution, we have

$$z + \ell\frac{\partial \ell}{\partial \ell} = \frac{\partial g_i}{\partial h_i}\frac{\partial \pi}{\partial \ell} + \frac{\partial f_1}{\partial q_1} q \tag{13.13b}$$

This is equation (A13.9f) in the appendix to this chapter. Recalling that $\frac{\partial \ell}{\partial \ell} < 0$, the left-hand side of (13.13b) is less than that of expression (13.12) in the previous subsection, so that recruiting will be curtailed if monopoly power is present, and the rate of recruiting will approach that of the intermediate case, from below, as $\frac{\partial \ell}{\partial \ell} \to 0$, that is, as monopoly power approaches zero. Define the elasticity of gross side payment z with respect to the rate of customer recruitment b as

$$\varepsilon = -\frac{z}{\ell}\frac{1}{\dfrac{\partial h}{\partial \ell}} \qquad (13.14)$$

then we have

$$z = \frac{\dfrac{\partial g_i}{\partial h_i}\dfrac{\partial z}{\partial \ell} - \dfrac{\partial f_\text{\tiny 1}}{\partial q_i}q}{1 - \dfrac{1}{\varepsilon}} \qquad (13.15)$$

This condition resembles the Marshall–Lerner condition of traditional economic theory, especially if we interpret the numerator as an expression of marginal cost (from the perspective of the interests of insiders). However, it has nothing to do with the *price* elasticity of demand. What this suggests is that the effects of monopoly power are to be found, not in the price at which the last unit of output is sold, but in the overall division of the surplus between insiders and new customers. To the extent that new customers are often given special conditions, more attractive than the terms available to insiders, we might expect that those conditions will be less favorable in the presence of greater monopoly power than otherwise.

No doubt noncooperative monopoly pricing, as portrayed in the neo-classical economic theory of monopoly, is observed in the actual world. Some instances were given in Chapter 11, section 11.3.1. This raises some issues of public policy that are far murkier than the traditional theory would make them seem. On the one hand, "market-based regulation" – regulations that create a market with a tendency towards a "law of one price" – may bring about a nearer approximation to a cooperative relation *between insiders and outsiders* than would otherwise occur (McCain, 2014a, Chapter 8). On the other hand, public enforcement of a law of one price may well exaggerate problems of monopoly power where they exist. For some instances of public policy failures as a consequence of neglecting this in the context of auction pricing, see Klemperer (2002).

Finally, noncooperative monopoly pricing might be observed because decision makers in the actual world do not always act with the sophisticated rationality assumed by the models in this chapter. They may not price cooperatively because they are not rational enough to cooperate.

We noted in the previous chapter that the commitment of more or less resources to search in labor markets would generate externalities through changes in the tightness of labor markets. Changes in the tightness of markets will also generate externalities in consumer markets. In the absence of monopoly power, increased marketing effort by sellers will generate positive externalities to potential customers, and negative externalities to other sellers, by shifting their costs of market participation. For a literal monopoly, however, the disadvantages of a tighter market are internal rather than external. Thus, there may be some gain in the efficiency of promotion for a monopoly relative to a more competitive group of firms. A more detailed investigation of this point will, however, be beyond the scope of this book.

13.3 CONCLUSION

In the *Principles of Political Economy* (1898/1965, p. 272) Mill writes "The monopolist can fix the value as high as he pleases, . . . but he can do so only by limiting the supply." The subsequent 150 years of monopoly theory have been little more than an elaboration of Mill's insight. If instead we understand the terms of exchange as the joint decision of *cooperating* buyers and sellers, Mill and all his followers are mistaken. Monopoly profits will not require a restriction of supply. Monopoly power will, however, increase the bargaining power of insiders as against outsider consumers. However, monopoly will not, in itself, be a source of misallocation of resources.

The theory of monopsony was developed by Robinson (1933, p. 223) as a straightforward counterpart to monopoly theory, and shares its shortcomings. (Robinson did, however, recognize the possibility of efficient price discrimination in monopsony markets, p. 227.) Here again, however, a cooperative perspective suggests that the influence of monopsony power will be primarily on the relative bargaining power of insiders and outsiders and not on the allocation of resources.

APPENDIX A13

This appendix considers the results of some modifications of the maximization program in the appendix to Chapter 11, to allow for decisions *by the continuing members as a coalition* to recruit new members, based on the expected parameters of representative new members. The main objective is comparison to some well-known results in neoclassical economic theory, including the consequences of monopoly and monopsony power. The appendix begins with recruitment in labor markets, and in each instance begin with an intermediate case in which there are costs of recruitment but no market power in the recruitment process.

To characterize the intermediate case in the recruitment of new employees, we have

$$L \leq \sum_{i \in E} k_i h_i + \ell - \imath(\ell) - L_0 \qquad (A13.1a)$$

Since

$$mz = \frac{\ell z}{hk} \qquad (A13.1b)$$

in place of (A11.3d) we will have

$$\Pi = \sum_{i \in B} z_i - \sum_{i \in E} z_i - \frac{\ell z}{hk} \qquad (A13.1c)$$

Here z is the wage of a representative new recruit, k the representative recruit's productivity, and h the resulting effort assignment. The Lagrangean function, in place of (A11.4b) will be

$$\mathcal{L} = \mathcal{Y} + \mu_1 \left(\sum_{i \in E} h_i k_i + \ell - \imath(\ell) - L_0 - L \right) + \mu_2 (F(L) - Q)$$

$$+ \mu_3 \left(Q - \sum_{i \in B} q_i \right) + \mu_4 \left(\sum_{i \in B} z_i - \sum_{i \in E} z_i - \frac{\ell z}{hk} - \Pi \right)$$

$$+ \mu_5 \left(1 - \sum_{i \in N} \lambda_i \right) \qquad (A13.2a)$$

Thus, in addition to the necessary conditions in the appendix to Chapter 11, we have

$$\frac{\partial \mathcal{L}}{\partial \ell} = \mu_1 \left(1 - \frac{\partial \imath}{\partial \ell} \right) - \mu_4 \frac{z}{hk} \leq 0 \qquad (A13.2b)$$

We cannot assume an interior solution in this case, since $\ell = 0$ is a possibility. For now, however, assume $\ell > 0$. Then, as before, we have

$$\mu_1 = \mu_3 \frac{\partial F}{\partial L} = \mu_4 \frac{\partial F_1}{\partial q_1} \frac{\partial F}{\partial L} \tag{A13.2c}$$

Thus

$$\frac{\partial f_1}{\partial q_1} \frac{\partial F}{\partial L} \left(1 - \frac{\partial z}{\partial \ell} \right) = \frac{z}{hk} \tag{A13.2d}$$

In case $\frac{\partial z}{\partial \ell} = 0$, we have a very familiar condition – recruitment is increased to the point at which the value marginal product (VMP) of effective labor is equal to the wage payment per unit of effective labor. Allowing for $\frac{\partial z}{\partial \ell} > 0$, we see that the VMP must be reduced by the marginal effort cost of recruiting a unit of effective labor. If, however, for $\ell > 0$, we have

$$\frac{\partial f_1}{\partial L_1} \frac{\partial F}{\partial L} \left(1 - \frac{\partial z}{\partial \ell} \right) = \frac{z}{hk} \tag{A13.2e}$$

then $\ell = 0$ is necessary. We have a corner solution and the coalition will not recruit.

To characterize monopsony power, we assume in addition that z increases with ℓ. Thus

$$z = h(\ell) \text{ with } \frac{\partial h}{\partial \ell} > 0. \tag{A13.3a}$$

Then, in place of (A13.2b) we have

$$\frac{\partial \mathcal{L}}{\partial \ell} = \mu_1 \left(1 - \frac{\partial z}{\partial \ell} \right) - \mu_4 \left(\frac{z}{hk} + \frac{\ell}{hk} \frac{\partial h}{\partial \ell} \right) \le 0 \tag{A13.3b}$$

using equation (A11.7g) in the appendix to Chapter 11, which is unchanged, we have

$$\frac{\partial f_1}{\partial q_1} \frac{\partial F}{\partial L} \left(1 - \frac{\partial z}{\partial \ell} \right) = \frac{z}{hk} + \frac{\ell}{hk} \frac{\partial h}{\partial \ell} \tag{A13.3c}$$

Comparison with (A13.2d) shows that monopsony power will lead the coalition to curtail its recruitment of new members at a level at which the marginal product of labor is greater than the wage cost, a result that formally resembles the neoclassical result on monopsony power.

To characterize the intermediate case of recruiting new customers with costs of recruiting but no monopoly power we have

$$L \leq \sum_{i \in E} k_i h_i - z(b) - L_0 \tag{A13.4a}$$

where b is the expected value of the number of representative new customers. Each representative customer takes q of the product. Thus, in place of (A11.1b),

$$\sum_{i \in B} q_1 \leq Q - bq \tag{A13.4b}$$

and in place of (A11.3d),

$$\Pi = \sum_{i \in B} z_1 + bz - \sum_{i \in E} z_i \tag{A13.4c}$$

The Lagrangean function then becomes

$$\mathcal{L} = y + \mu_1 \left(\sum_{i \in E} h_i k_i - z(b) - L_0 - L \right) + \mu_2 (F(L) - Q)$$

$$+ \mu_3 \left(Q - \sum_{i \in B} q_1 - bq \right) + \mu_4 \left(\sum_{i \in B} z_1 + bz - \sum_{i \in E} z_i - \Pi \right)$$

$$+ \mu_5 \left(1 - \sum_{i \in N} \lambda_i \right) \tag{A13.5}$$

Among the necessary conditions is

$$\frac{\partial \mathcal{L}}{\partial b} = -\mu_1 \frac{\partial z}{\partial b} - \mu_3 q + \mu_4 z \leq 0 \tag{A13.6a}$$

Once again we may not assume that this is an interior solution. In case it is, however, we have equivalently

$$z = \frac{\partial g_i}{\partial h_i} \frac{\partial z}{\partial b} + \frac{\partial f_1}{\partial q_1} q \tag{A13.6b}$$

(using A11.7h and A11.7j, which are unchanged). In case

$$z < \frac{\partial g_i}{\partial h_i} \frac{\partial z}{\partial b} + \frac{\partial f_1}{\partial q_i} q \tag{A13.6c}$$

for $b \geq 0$, we have a corner solution at $b- = 0$: the coalition will not recruit new customers.

In the case of positive monopoly power, we have, in addition, the constraint

$$z = z(b) \text{ with } \frac{\partial z}{\partial b} < 0 \tag{A13.7}$$

Thus, the Lagrangean function becomes, in place of (A13.5)

$$\mathcal{L} = y + \mu_1\left(\sum_{i \in E} h_i k_i - r(\ell) - L_0 - L\right) + \mu_2(F(L) - Q)$$

$$+ \mu_3\left(Q - \sum_{i \in B} q_i - \ell q\right) + \mu_4\left(\sum_{i \in B} z_i + \ell z - \sum_{i \in E} z_i - \Pi\right)$$

$$+ \mu_5\left(1 - \sum_{i \in N} \lambda_i\right) + \mu_6(k(\ell) - z) \tag{A13.8}$$

Then, in place of (A13.6a) we have

$$\frac{\partial \mathcal{L}}{\partial \ell} = -\mu_1 \frac{\partial r}{\partial \ell} - \mu_3 q + \mu_4 z + \mu_6 \frac{\partial k}{\partial \ell} \leq 0 \tag{A13.9a}$$

$$\frac{\partial \mathcal{L}}{\partial z} = \mu_4 \ell - \mu_6 \leq 0 \tag{A13.9b}$$

with an interior solution on z we have

$$\mu_6 = \mu_4 \ell \tag{A13.9c}$$

so that (A13.9a) becomes

$$\frac{\partial \mathcal{L}}{\partial \ell} = -\mu_1 \frac{\partial r}{\partial \ell} - \mu_2 q + \mu_4\left(z + \ell \frac{\partial k}{\partial \ell}\right) \leq 0 \tag{A13.9d}$$

Once again, using (A11.7h) and (A11.7j), with an interior solution for b,

$$\mu_4\left(z + \ell \frac{\partial k}{\partial \ell}\right) = \mu_4 \frac{\partial g_i}{\partial h_i}\frac{\partial r}{\partial \ell} + \mu_4 \frac{\partial f_i}{\partial q_i} q \tag{A13.9e}$$

$$z + \ell \frac{\partial k}{\partial \ell} = \frac{\partial g_i}{\partial h_i}\frac{\partial r}{\partial \ell} + \frac{\partial f_i}{\partial q_i} q \tag{A13.9f}$$

If, however, for $b > 0$, we have

$$z + \ell \frac{\partial k}{\partial \ell} < \frac{\partial g_i}{\partial h_i}\frac{\partial r}{\partial \ell} + \frac{\partial f_i}{\partial q_i} q \tag{A13.10}$$

then $b = 0$ is necessary: we have a corner solution at which new customers will not be sought. In (A13.9f) and (A13.6b) we can construe the left-hand side as the marginal benefit of recruiting and the right-hand side as the marginal cost of recruiting from the point of view of insider employees (first term) and customers (second term). Thus, these expressions recapitulate the familiar idea that an activity is carried on until its marginal cost is equal to its marginal benefit. Then comparing (A13.9f) and (A13.6b) we see that, for otherwise comparable circumstances, the marginal benefit of recruiting is less when monopoly power is present, and approaches the intermediate case, from below, as $\frac{\partial \dot{a}}{\partial \dot{q}} \rightarrow 0$, that is, as monopoly power approaches zero.

14. Bargaining and the determination of wages

This part of the book has explored the view of the business firm as a cooperative coalition. Since the costs of coalition formation are sunk, and the same costs limit the formation of competitive firms, the successful coalitions we observe typically will realize surpluses. As a result bargaining power plays a key part in the models of Chapters 11 and 13. In these models the wage includes a share of the surplus and this is necessarily positive, since otherwise employees would not recover the sunk cost of their participation in labor markets. Thus, even in the absence of explicit collective bargaining, a labor share of the surplus must be determined, and the determinants of that share constitute (for the purposes of this book) bargaining, implicit if not explicit. Accordingly, this chapter reconsiders bargaining power and the determination of wages.

14.1 THREATS AND THEORIES OF BARGAINING

Bargaining power arises from threats. In a successful cooperative arrangement the threats are not carried out, so are not necessarily observable. Instead rational bargainers each assess the threats that others may make and moderate their demands so that a mutual agreement can occur. When we observe threats carried out, the attempt to form a cooperative coalition or to make a cooperative decision has failed. Most bargaining theory in the tradition of Zeuthen (1930) and Nash (1950b) explicitly consider only the threat to leave the coalition. For such a threat, the disagreement outcome – that is, the outcome if the threat is carried out – is the bargainer's outside option. In Schmeidler's (1969) nucleolus model, similarly, bargaining power depends on the difference between the payoff to a group or a singleton within the coalition and the value they can attain in another cooperative grouping, that is, the outside option. This identification of the disagreement outcome with the outside option is common in applications of bargaining theory, but not the only possibility. Nash (1953) relaxes this assumption somewhat, but Nash seems to assume that if there is no bargaining settlement there is no coalition.

In Nash's variable threat model, the noncooperative choice of threat strategies determines the outcomes in case bargaining fails. Nash does not provide examples. The widely discussed efficiency wage model may provide a useful instance. It is not a bargaining model, as usually presented, but rather a Stackelberg leader–follower model with the employer as the leader. Nevertheless, we may identify a noncooperative solution and a cooperative solution that might be maintained with repeated play (MacLeod and Malcomson, 1989). At the noncooperative solution, effort and productivity are low, and pay is at the market-clearing rate. At the cooperative solution, effort and productivity are higher and so is the wage. The employees' threat in this case is to withdraw effort to the noncooperative level. The employer's threat is to cut pay. In case the threats are carried out (and the employees, as individuals, choose not to carry out the threat to quit) the coalition is not dissolved. Instead it continues with (noncooperative) reduced effort and low pay. For the employer, at least, this disagreement outcome is not dependent on the outside option.

Indeed, actual bargaining clearly sometimes takes place on the basis of threats that do not dissolve the coalition. Strikes and lockouts provide well-known examples. When employees go on strike, or when hockey players are locked out and part of their play schedule is cancelled, neither side means to leave the coalition or the league permanently. Another illustration of this sort of threat is the work-to-rules slowdown. If a work-to-rules slowdown is carried out, the coalition is not dissolved, and indeed production continues, but the rate of production is reduced and the size of the surplus produced by the coalition is diminished. The disagreement outcome depends on how the reduction in the surplus is allocated among the members of the coalition, presumably a noncooperative solution of some kind. In case one member receives the residual, while others are paid according to a pre-existing contract, the threat of the slowdown may enhance the bargaining power of those seeking a new contract.

This suggests one possible explanation of the failure of the bargaining model to fit the data on the (lack of) fluctuation of wages with unemployment: the bargainers might choose threats with disagreement outcomes that are independent of the outside options of the bargainers. In effect this is the strategy of Hall and Milgrom (2008). They adopt a different bargaining theory along the lines of Rubinstein (1982) in which offers are exchanged in a subgame perfect sequence. In that case, the disagreement outcomes are delays that are costly to both bargainers (note also Cross, 1965). However, this approach seems applicable only to new hires, and would be applicable to the wages of insiders only to the extent that insiders' wages are determined for all time by the wage at the initial hire.

A shortcoming of Nash's variable-threat model is that a bargainer, having chosen one threat, must renounce all others, so that only one threat influences the bargaining process. (He does assume, however, that the first-stage equilibrium will be a mixed strategy equilibrium, so that the ultimate outcome of the process may be influenced probabilistically by two or more threats.) But why would a bargainer choose just one among a number of threats, rather than use any and all threats that might extract a concession from the other side, at any stage in the bargaining process?

Threats may be classified by noncooperative credibility: those that, in Nash's (1953) words, "will not be something [the threatener] would want to do, just for itself" (that is, that are not subgame perfect; Chapter 6 above) and those that are. In a cooperative game in which the existing allocation is dominated, the threat to leave the coalition is credible; otherwise not. In this way the core of a cooperative game may be thought of as a bargaining theory, although one that does not always give rise to a unique value solution. On the other hand, if the allocation is not dominated, the threat to leave the coalition will never be subgame perfect. Thus Nash says that "we must assume there is an adequate mechanism for forcing the players to stick to their threats and demands once made" (p. 130). But Nash goes too far, here. Strikes and lockouts are surely almost never subgame perfect, but they do occur in the actual world. It seems clear that real human beings have some capacity to carry out threats of retaliation, some capacity for spiteful (*New York Times*, 2014) behavior. That is, bargainers' behavior may be more closely approximate ideal than perfect rationality.

Multiple threats, including some that are not credible in a world of "perfect" rationality and subgame perfect equilibrium, may explain differences in bargaining power that are independent of the outside options of the bargainers. Differences in bargaining power along these lines are allowed for by generalizations on Nash's bargaining theory due to Roth (1979) and Svejnar (1986). These generalizations also allow for bargaining among more than two bargainers, but assumes that only individuals have bargaining power.

14.2 BARGAINING POWER AND COLLECTIVE ACTION

In this part of the book so far, it has been assumed (along the lines of Roth, 1979 and Svejnar, 1986) that bargaining power may differ from one individual to another, but only individuals have bargaining power. But the history of human action points to a possibility excluded by that assumption: individuals may often enhance their bargaining power by banding together. The examples of strikes and slowdowns already discussed would seem to be instances of this. If so, it would seem that this possibility is particularly important in the modern business firm. With this possibility in mind, it will be appropriate to reconsider the role of collective action in determining bargaining power and the results of the study so far.

Suppose, then, that employees gain some bargaining power as a group by means of a threat of collective action such as a strike, work-to-rules slowdown, or other noncooperative effort withdrawal. A model along these lines is developed in McCain (2013, Chapter 6, Section 6.2) for a TU game. For the NTU game discussed in the appendix to Chapter 11 in connection with expressions (A11.1a)–(A11.7n), the gain for the workforce as a whole cannot be expressed except in terms of the distributive weights λ_i. Accordingly, adapting the bargaining model of McCain (2013, Chapter 6, section 6.5) to this case, the objective function for the maximization program is as shown in the appendix to this chapter as (A14.1a). A new term for the aggregate gain of the employee group is added to the objective function of the decisions.

For non-employee members of the coalition, we see

$$\lambda_j = \frac{\beta_j}{\beta_E + \sum_{k \in N} \beta_k} \tag{14.1}$$

where β_E is the bargaining power of the employees as a group. That is, for non-employee members of the coalition, as before, the distributive weight λ_i will be the proportion of the individual's bargaining power to the total bargaining power in the game. This is equation (A14.1d) in the appendix. In this case the total bargaining power includes that of the employee group as a whole, β_E. For employees the case is more complex, but the distributive weight λ_i will be increased by a proportion of the collective bargaining power β_E. This is equation (A14.1e). Thus, the threat of collective action reduces the relative bargaining power of those outside the group that threaten the collective action, as we might expect. Nevertheless, the firm's allocative decisions will not be modified by this shift in bargaining power (equations A14.4a–c) and the maximized net value creation will be distributed among the members of the coalition in proportion to their distributive weights as modified by the threat of collective action. That is to say, briefly, the impact of a threat of collective action will be only on the distribution of the surplus, but will not affect the coalition's decisions on the allocation of resources nor, consequently, on the surplus value itself. The value creation approach is applicable despite the exercise of collective bargaining power.

In this model (following Schmeidler, 1969 and McCain, 2013) the bargaining model is treated as a core assignment algorithm. That is, it distributes the entire surplus of the coalition, not the excess over some threat outcome. The role of threats is to determine bargaining power. Since the surplus is defined relative to the outside option, the outside option will enter into the determination of the wage, z_i.

14.3 WAGE AND UNEMPLOYMENT

Accordingly, we return to the relation of unemployment to the bargaining-determined wage. How much, indeed, would we expect the wage z_i to vary with changes in the rate of unemployment? Recall that z_i has three components (equation A11.9k in the appendix to Chapter 11 and A14.3b in the appendix to this chapter):

$$z_i = \lambda_i V + g_i(h_i) + w_i \qquad (14.2)$$

For an employee whose second-best alternative is to seek other employment, supposing the search for new employment is successful, the payoff of the new job will have the same components. Moreover, if the job search fails in the current period, the individual may obtain employment in a future period. Similarly, if he remains in his present employment, he can expect remuneration in future periods, which will comprise the same three components, along with some probability of disemployment in the future periods. These sequences of expected future remuneration could be quite complex, but we are concerned here with the response to changes in unemployment in the *current* period, so we let the expected discounted value of remuneration in future periods, conditional on separation from the current job, be a constant Z^1. We let the expected discounted value of remuneration in future periods conditional on continuation in the job for the current period be a constant Z^2. Let the probability of re-employment in the current period be p, y^* the net share of the surplus to be expected in that case, and g^* the expected value of the compensation for effort in the new job. Suppose further that there is a fixed cost, γ, of entering the labor market. This might be a resource cost such as the cost of travel to attend interviews. It might also be a subjective fixed cost. Surveys of self-assessed quality of life indicate that the unemployed are less happy than those who are employed *at the same income* (for example, Frey, 2008). Then this subjective fixed cost is a component of γ. Thus, a rational agent then will leave the job if

$$y_i + Z^2 < py^* + Z^1 - \gamma \qquad (14.3)$$

That is, the rationality constraint $y_i \geq w_i$ is equivalent to

$$y_i \geq py^* + Z^1 - Z^2 - \gamma \qquad (14.4)$$

That is,

$$w_i = py^* + Z^1 - Z^2 - \gamma = p(z^* - g^* - w_i) + Z^1 - Z^2 - \gamma \quad (14.5)$$

Assume for simplicity that $Z^1 = Z^2$. Then (14.5) becomes

$$(1 + p)w_i = p(z^* - g^*) - \gamma \qquad (14.6)$$

That is,

$$w_i = \frac{p}{1 + p}(z^* - g^*) - \frac{\gamma}{1 + p} \qquad (14.7)$$

Thus,

$$z_i = \lambda_i V + g_i(h_i) + \frac{p}{1 + p}(z^* - g^*) - \frac{\gamma}{1 + p} \qquad (14.8)$$

With V the coalitional surplus in the present job net of the outside options of all members of the coalition. This will be illustrated by a numerical example. For simplicity and for comparability with received theory, assume that all employees have the same outside option $y^* = z^* - g^* - w_i$, $g_i(h_i) = g^*$, and that there are n employees. Then

$$z_i = \lambda_i \left(V^* - w_A - \sum_{i \in B} w_i \right) + (1 - \lambda_i n) \left(\frac{p}{1 + p} z^* - \frac{\gamma}{1 + p} \right)$$
$$+ \left(1 - \frac{p}{1 + p} (1 - \lambda_i n) \right) g^* \qquad (14.9)$$

As p varies from zero to one $\frac{p}{1+p}$ varies from 0 to 0.5 and $\frac{1}{1+p}$ varies from 1 to 0.5. Thus, the first term is a constant with respect to p, the last varies directly with p, and the middle term varies inversely with p, provided $pz^* > \gamma$.

As a first approximation, let the overall average p be the ratio of the number of job openings, J, to the number unemployed, U. This may oversimplify – some job openings will be filled directly by transfers from other jobs, without a period of unemployment, and some openings may indeed not be available to the unemployed, for example. Further, Chapter 12 has proposed a different measure. Nevertheless, for simplicity and comparability with the received literature, and preliminarily, posit

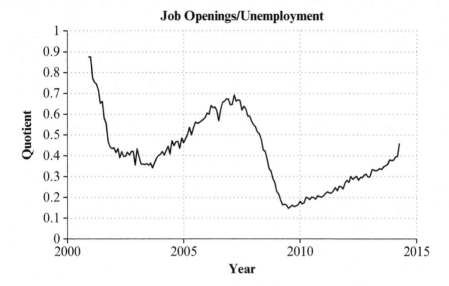

Figure 14.1   Job openings as a proportion of unemployment

$$p = \frac{J}{U} \tag{14.10}$$

Taking data on the level of job openings in the United States for 2001–2014 from the JOLTS project of the Bureau of Labor Statistics and data on unemployment levels from the Bureau of Labor Statistics for the same period, we see the quotient as plotted in Figure 14.1. On the one hand, the quotient varies between 0 and 1 and so may, without contradiction (at least in this period) be interpreted as a probability. On the other hand, it varies over most of that range in a highly countercyclical way. Thus it would seem that, by identifying this quotient with the probability of quick re-employment, we err in the direction of exaggerating the impact of cyclical variations on the computed wage, rather than in the opposite direction.

For a numerical example, suppose that $\lambda_i n = 1/2$, $z^* = 42{,}500$, $g^* = 35{,}000$, and $\lambda_i(V^* - w_A - \Sigma_{i \in B} w_i) = 1000$ and $\gamma = 5000$. Note that $\lambda_i n = 0.5$, would imply that employees as a group obtain about half of the surplus: if λ_i is the same for all employees, then they receive exactly half. Then we obtain the calculated values for z_i in four contrasting periods as shown in Table 14.1.

We see that the projected value of z varies only slightly as the unemployment rate fluctuates from its lows to its highs over "the great recession." The variation is less than 2 percent. The quotient J/U and the predicted

Table 14.1   Computed wage payments

Period	Average unemployment rate	Average quotient	Computed wage
2002–2005	5.6	0.439459449	46144.85541
2006–2008	5.008	0.561981044	46349.20262
2009–2011	9.28	0.200062587	45625.16298
2010–2014	7.6	0.331186084	45932.96334

Quotient and Computed Wage

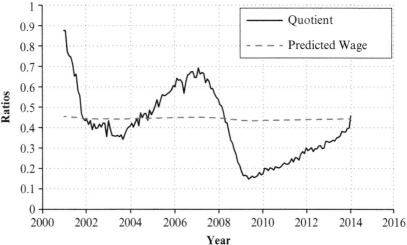

Figure 14.2   The predicted wage and the quotient J/U

wage are plotted together in Figure 14.2 with the predicted wage expressed as a proportion of 100,000 for comparability.

A reason for this result is that g^*, the compensation for the disutility of effort, is relatively large by comparison with z^*, the expected wage conditional on re-employment, and z_i. Since the individual's bargaining share, y_i, is *net* of this compensation, and the compensation recurs in the case of re-employment, it reduces the impact of J/U on the predicted wage. To illustrate this, consider a second example with $z^* = 40,000$, $g^* = 0$, $\gamma = 0$, and $\lambda_i(V^* - w_A - \Sigma_{i \in B} w_i) = 40,000$. Then we have Table 14.2 and Figure 14.3 by contrast with Table 14.1 and Figure 14.2. While the variation in the wage from the low to high points in 2002–2014 does not appear large by comparison with that of J/U, it is in the range of 9 percent, about five times as great as the previous example.

Table 14.2   Computed wage payments in an alternative example

Period	Average unemployment rate	Average quotient	Computed wage
2002–2005	5.6	0.439459449	46105.89551
2006–2008	5.008	0.561981044	47195.7473
2009–2011	9.28	0.200062587	43334.20255
2010–2014	7.6	0.331186084	44975.80448

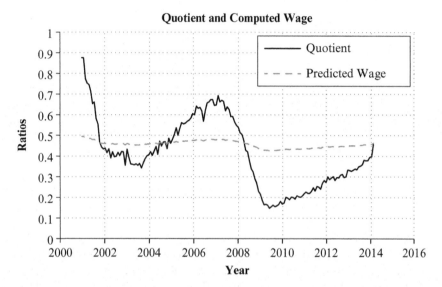

Figure 14.3   The predicted wage and the quotient J/U in the alternative example

It appears that the application of bargaining theory to the determination of wages in search-and-matching models has overlooked a key question: what proportion of the wage is compensation for the disutility of effort? If that proportion is large, then the model does not predict any considerable countercyclical variation of wages with unemployment. We note that the compensation for effort depends on the structure of production, and not on the employee's outside option. In a more complex model, there might be other components of employee compensation that depend on the persistent structure of production rather than on the employees' outside option, and this result probably is more valuable as an instance of that dependence than as a complete theory of wages. The question should be, to what extent are wages determined by the structure of production

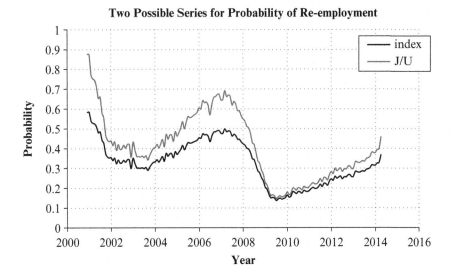

Figure 14.4   Two measures of the probability of re-employment

rather than by the employee's outside option? The evidence that real wages
are sticky suggests that wages are primarily determined by the persistent
structure of production.

All in all, the prediction that the wage would fluctuate countercyclically
is weakly supported if at all. If the bargaining model is somewhat more
realistically complex, and in particular includes a component of the wage
that compensates for efficient effort, then the predicted countercyclical
fluctuation in wages could be very slight, if not indeed negligible.

Thus far we have used the ratio J/U as the probability of prompt re-
employment in determining the individual employee's outside option and
thus wage. Suppose instead we adopt the index $p = [1 - (1 - \frac{1}{U})^J]$ suggested by the search model of section 12.2.1. A preliminary examination of
the two probability expressions suggests that there will be little difference,
but that this substitution would further reduce the responsiveness of the
wage to unemployment. This is suggested by Figure 14.4, which compares
the two estimates of the probability of re-employment.

Then, replacing *J/U* with the index, we have Table 14.3 in place of
Table 14.1, and in place of Figure 14.2 we have Figure 14.5.

It seems that the bargaining hypothesis cannot be dismissed on the
grounds that the wage does not vary proportionately with the rate of
unemployment.

Table 14.3   Computed wage payments

Period	Average unemployment rate	Average index	Computed wage
2002–2005	5.6	0.354234575	45980.90809
2006–2008	5.008	0.425803662	46119.90436
2009–2011	9.28	0.180904544	45574.46814
2010–2014	7.6	0.281289003	45823.25982

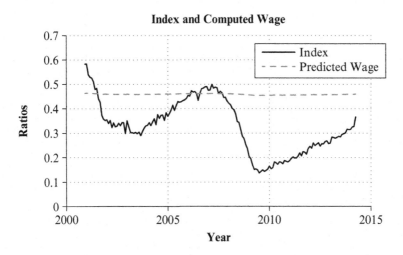

Figure 14.5   The probability of re-employment and the predicted wage in a third example

14.4   SUMMARY AND CONCLUSION

This chapter has revisited two aspects of models in the previous chapters. First, the threat of some collective action by a group of members of the coalition was introduced into the bargaining model that defined a considerable solution in Chapter 11. The group threatening collective action is assumed to comprise the employees. The implication of such a model is that threats of collective action shift the relative bargaining power of those within and without the threat group, but do not influence the allocation of resources (maintaining the normalization assumptions of this part of the book) so that the TU game analysis may be employed.

The chapter then revisited the search-and-matching model from Chapter 12 section 12.2 to explore the implications of the model for covariance of unemployment and real wages. It is found that – when we take into account some complexities often neglected in the literature – the covariance would be very weak, as the empirical evidence suggests.

Both of these generalizations point up the importance of distinguishing the situations of insiders and outsiders in a cooperative-game model of the firm. Outsiders cannot influence bargaining through threats, and the relations between outsiders' expectations and insiders' bargaining power seems far less direct than is often assumed.

APPENDIX A14

The model of Chapter 11 will now be modified to allow for group bargaining power based on a threat of collective action by the employees. The objective function (A11.4a) becomes

$$y = \sum_{i \in B} \beta_i \ln(\lambda_i y_i) + \sum_{i \in E} \beta_i \ln(\lambda_i y_i) + \beta_A \ln(\lambda_A y_A)$$

$$+ \beta_E \ln\left(\sum_{j \in E} \lambda_j y_j\right) \tag{A14.1a}$$

Where β_E is the bargaining power of employees as a group, arising from their collective threat of noncooperation. In place of (A11.5b) we have

$$\forall i \in E, \frac{\partial \mathcal{L}}{\partial \lambda_i} = \frac{\beta_i}{\lambda_i} + \beta_E \frac{y_i}{\sum\limits_{k \in E} \lambda_k y_k} - \mu_5 \leq 0 \tag{A14.1b}$$

As before, and for the same reasons, we consider only interior solutions. Noting that (A11.5a) and (A11.5c) are unchanged, we have, in place of (A11.5d),

$$\mu_5 = \beta_E + \sum_{k \in N} \beta_k \tag{A14.1c}$$

and in place of (A11.5d) we have

$$\forall j \in N|E, \lambda_j = \frac{\beta_j}{\beta_E + \sum\limits_{k \in N} \beta_k} \tag{A14.1d}$$

Thus, as before, for those outside the group that poses the threat of collective action, the distributive multiplier is the individual's proportion of the total bargaining power in the game, but that proportion is diminished by the addition of the group bargaining power β_E. Further,

$$\forall i \in E, \lambda_i = \frac{\beta_i + \beta_E \dfrac{\lambda_i y_i}{\sum\limits_{k \in E} \lambda_k y_k}}{\beta_E + \sum\limits_{k \in N} \beta_k} \tag{A14.1e}$$

It must be noted that this complex equation does not determine λ_i, since λ_i is on both sides of the equation. To determine λ_i we would require a still more complex polynomial. However, (A14.1e) lends itself to ordinary-language interpretation: relative to nonmembers of the group, the distributive multiplier for members of the threat group is incremented relative to individual bargaining power by a factor that depends on the individual share of the group's bargaining power.

Expression (A11.6a) becomes

$$\forall i \in E, \frac{\partial \mathcal{L}}{\partial z_i} = \frac{\beta_i}{y_i} + \beta_E \frac{\lambda_i}{\sum_{k \in E} \lambda_k y_k} - \mu_4 \le 0 \qquad \text{(A14.2a)}$$

$$\forall \iota \in B, \frac{\partial \mathcal{L}}{\partial z_\iota} = -\frac{\beta_\iota}{y_\iota} + \mu_4 \le 0 \qquad \text{(A14.2b)}$$

$$\frac{\partial \mathcal{L}}{\partial z_A} = \frac{\beta_A}{y_A} - \mu_4 \le 0 \qquad \text{(A14.2c)}$$

Thus, in place of (A11.8e) we have

$$\mu_4 = \frac{\beta_E + \sum_{k \in N} \beta_k}{V} \qquad \text{(A14.2d)}$$

$$\forall j \in N|E, y_j = \frac{\beta_j}{\beta_E + \sum_{k \in N} \beta_k} V = \lambda_i V \qquad \text{(A14.3a)}$$

$$\forall i \in E, y_i = \frac{\beta_i + \beta_E \dfrac{\lambda_i y_i}{\sum_{k \in E} \lambda_k y_k}}{\beta_E + \sum_{k \in N} \beta_k} V = \lambda_i V \qquad \text{(A14.3b)}$$

Once again, while λ_i and y_i are not determined by this complex expression, we see that the distribution is shifted toward members of the group that pose the threat of collective action and away from the others, relative to their individual bargaining power.

In place of (A11.7c) we have

$$\forall i \in E, \frac{\partial \mathcal{L}}{\partial h_i} = -\frac{\partial g_i}{\partial h_i} \left(\frac{\beta_i}{y_i} + \frac{\beta_E}{\sum_{k \in E} \lambda_k y_k} \lambda_i \right) + \mu_1 k_i \le 0 \quad \text{(A11.4a)}$$

Using (A14.2a), that is

$$\forall i \in E, \frac{\partial \mathcal{L}}{\partial h_i} = -\frac{\partial g_i}{\partial h_i} \mu_4 + \mu_1 k_i \leq 0 \qquad \text{(A14.4b)}$$

Thus, as before, despite the different determination of μ_4, we obtain, in parallel with (A11.7k),

$$\forall i \in E, \forall \iota \in B, \frac{\partial g_i}{\partial h_i} = \frac{\partial f_\iota}{\partial q_\iota} \frac{\partial F}{\partial L} k_i \qquad \text{(A14.4c)}$$

Thus, as before, the allocation of resources is determined by Paretian efficiency, and the effect of threats of collective action by the employees is limited to a shift in the distribution of the value created by the enterprise.

15. Bargaining power and majority rule

We often hear that in real-world politics, "wheeling and dealing" and exchanging one favor for another are as common as straightforward voting on the basis of "the issues," if not more common. On the other hand, some bitter experiences of the last century and this suggest that voting bodies that are unable to compromise are not likely to succeed as bodies of democratic governance. Compromise is a process in which each side gives up what it wants less in order to obtain what it wants more, exchanging one favor for another. Economists tend to believe that, at least in some circumstances, wheeling and dealing tends to improve efficiency.

It seems that decisions in many voting bodies might be described by a two-stage decision process in which the first stage is a bargaining process and the second is a vote that is often a formality. This does not mean that the voting is irrelevant, but, rather, that it limits the threats that may be made and so influences bargaining power at the first stage. This chapter will explore a two-stage game in which the first stage is a bargaining process and the game terminates if there is an agreement, while at the second stage, if there is no agreement at the first stage, a contested election is held to determine the joint strategy of the body. Bargaining power at the first stage is attributed to minimum winning coalitions in the possible second stage election.

In an idealization of such a two-stage game, majority groups have equal bargaining power, and nonmajority groups have none. In this chapter, as in the previous chapter, the analysis relies on a recent extension of bargaining theory that attributes bargaining power to groups as well as individuals. It is then assumed that a minimum winning voting block has bargaining power one and other groups and individuals have bargaining power zero. For TU games, this yields a striking rule for the bargaining solution: the surplus generated by the coalition is either distributed as equal payouts, or distributed among the members with lesser individual disagreement points, so that their payouts are equal, while others get their individual disagreement points.

15.1 NONCOOPERATIVE AND COOPERATIVE MODELS OF MAJORITY RULE

Majoritarian mechanisms for collective decisions have been studied by mathematicians, social philosophers, economists and game theorists for more than two hundred years. In addition to the long and distinguished literature on voting that extends from the Marquis de Condorcet to Howard Bowen (1943), we have game theoretic models of voting both in terms of cooperative and noncooperative games.[1] In the terms of game theory, many of these studies can be characterized as noncooperative models of voting. Noncooperative models treat the individual's decision to vote for one alternative or another as a noncooperative decision; that is, a decision based on the self-interest of the individual decision maker, who supposes that the decisions of all other individual decision makers are given and similarly based on their self-interest. This literature has tended to focus on the possibility of strategic voting. When an individual votes for a second or lower preference in the hope of obtaining a better outcome than a vote for the first preference would realize, this is *strategic voting*. Otherwise the vote is *naïve*. In the perspective of the theory of mechanism design, strategic voting is a problem. If a mechanism of information revelation will lead individuals to reveal their "types," then the mechanism can implement any cooperative objective for the group decision.[2] If the cooperative objective is efficiency, then we say that the mechanism is *incentive compatible*. In the presence of strategic voting, however, the individual's type is not reliably revealed. This has been shown by "reverse engineering" the Arrow impossibility theorem. Arrow showed that there can be no mechanism – voting, markets, or any other – that consistently aggregates preferences in general. If there were a voting rule that is robust against strategic voting, it would aggregate preferences consistently, so there cannot be such a rule. (Chapter 7, section 7.3.2).

There are two further difficulties in a noncooperative view of voting, one minor and one major. The minor one is that "naïve" voting may not reveal the individual's type, if types differ also in second and third or lower preferences. There are proposals for voting rules that allow the revelation of second and lower preferences, but these, too, are subject to the reverse-engineered Arrow principle. The major difficulty is that, if voting is not compulsory, noncooperatively rational decision makers will choose not to vote. Nevertheless, many people do choose to vote, and this seems a common instance of spontaneous, unforced cooperative action.

We may then turn to a cooperative model of voting. One approach would be to treat voting blocks as cooperative coalitions (see, for example, Anesi and de Donder, 2013). Indeed, such blocks are common and important enough in politics that the term "coalition," that we use for a cooperative grouping, is itself derived from European politics. The difficulty of a cooperative model of this kind is that, in general, it has no stable solutions. Consider, for example, a simple three-person game among Andrea, Bill, and Carol. Any coalition of two of them can divide a payoff of 1 between them in any proportion that they choose. Suppose, then, Andrea and Carol form a winning coalition and divide the payoff equally.[3] Then Bill can approach Andrea and offer to form a new winning coalition in which he will settle for one-third. This will make both Bill and Andrea better off. In this way, every winning coalition in a majority game is dominated by another winning coalition. This is called a *dominance cycle*. But, again, both of these views overlook two things. First, if people are rational, they will anticipate this dominance cycle and not go pursuing gains that will never be realized. Second, in all of this we are supposing that voting is a way of settling differences within a coalition already formed. The organization of an election is itself a cooperative act, with a set of commonly agreed rules and procedures: this presupposes an existing coalition of all the voters. Why then would such a coalition choose majority voting as a mechanism for its joint decisions?

In a world of free information and perfect rationality, the pre-existing coalition would decide its joint activity and the distribution of the benefits by mutual agreement, that is, unanimity rule. In a world of costly information and bounded rationality, however, majority decisions may be arrived at with much less costly information, so, imperfect as they may be, majoritarian decisions may yet be preferable to unanimity rule – supposing, indeed, that unanimous decisions are possible at all. In this elections are somewhat parallel to auctions (McCain, 2014a, pp. 123–4). Auctions have almost always been studied in terms of noncooperative decisions, but, in fact, the establishment of an auction is a cooperative act in itself. If information is costly, but each participant has costless knowledge of the value to herself of the items sold, then the auction is a very cheap, incentive-compatible mechanism. A cooperative group might well choose such a mechanism to reduce the informational costs of collective decision making. However, the auction mechanism also shifts bargaining power: those who are not efficient buyers have no bargaining power whatever in the allocation of the benefits of the auction. We may therefore ask, in parallel, what are the implications of majority rule for bargaining power? Is it possible that the pre-existing cooperative coalition might choose majority rule for some decisions because of the way it influences bargaining power?

15.2 BARGAINING POWER AND MAJORITY RULE

Bargaining power, we recall, arises from threats. If the disagreement outcome is that a contested election will be held, then there is a relatively limited range of threats that can be made. One can threaten to vote against an alternative that another participant very much wants, even if the threatener would prefer to vote for it, but one cannot (within the election game) threaten to beat him up or put a bomb in his mailbox. This will have an effect on the bargaining power within the coalition, and a postulate of this chapter is that elections are often chosen as a decision mechanism for just that reason, although the limited range of threats seems a good thing in itself.

This chapter of the book explores a two-stage collective decision process in which the first stage is a bargaining game and at the second stage, if no bargain is struck, a contested election is held to determine the joint strategy of the coalition. Bargaining power at the first stage is attributed to minimum winning coalitions in the possible second stage election. In the tradition of cooperative game theory, we assume that the bargaining is successful and explicitly consider only the bargaining stage.

15.2.1 Bargaining Power Games and Voting Games

This discussion relies on a formal definition of voting games as a particular category of bargaining power games, which in turn are characterized in McCain (2013, Chapter 6, section 6.1). In ordinary-language terms, a bargaining power game is a group of $n > 1$ decision makers who make joint use of a limited resource in ways that may vary over a continuum of possible methods. The methods are not represented in detail – that will be left to engineers and operations researchers – but is summarized by a constraint function, expression II in the appendix to this chapter. The constraint function expresses the difference between the resources needed to generate a set of payoffs y_i to the decision makers $i = 1, \ldots, n$ and the resources available. This excess resource requirement cannot be positive: this is the limit within which the group must function. If any subgroup decide to drop out of the agreement and operate separately, some of the resources may be available to the separate group, so that a constraint function is defined also for each subgroup. If the subgroup has no resources then the constraint function is positive for any positive payoff y_i for any member of the group (payoffs are nonnegative). Consequently, in that case, payoffs in the separate subgroup can only be zero. For an individual, the constraint function is the individual's payoff in case there is no agreement. Each subgroup is characterized by a bargaining power index, a constant β_S. This is expression III in the appendix to this chapter. In this chapter we will consider two special cases: transferable utility (TU) games, in which the unweighted sum of all payoffs is at most a constant, and linear games, in which a weighted sum of the payoffs is a constant.

The bargaining power game will be a voting game if, in addition, the subgroups can be classified into two types, so that all those who are of the first type have bargaining power $\beta_S = 1$, and for the second type, $\beta_S = 0$. The first type are winning voting blocks. In a majority voting game, groups of the first type are groups that comprise a majority. Now, the bargaining power game can be thought of as an algorithm to derive the index of individual bargaining power from the collective bargaining power of the groups in which the individual may participate. (The previous chapter supplies an instance.) Thus, groups of the first type are *minimum* winning voting blocks, so that each participant in the voting block is essential to its majority. This is expression IV.ii in the appendix to this chapter.

An important property of majoritarian processes is that majorities are *symmetrical*. That is, if S is a minimal winning block and one of its members drops out, then that voter can be replaced by any former non-member. The new block generated in this way is also a minimal winning voting block. Defining a majority rule voting game in the obvious way, that winning blocks are blocks with no less than $(n + 1)/2$ members, it is clear that majority rule is symmetrical. This is Lemma 3 in the appendix to this chapter.

Now consider a mechanism like that of the United States Senate: there are an even number of ordinary members plus a chairperson who votes only to break a tie. Is this mechanism symmetrical? In principle (that is, ignoring political parties) it is. A minimum winning voting block is a block of 50 percent + 1, and this is so regardless whether the chairperson votes or not. The chairperson does not derive any special disadvantage or advantage from the fact that her vote only "counts" if it is required to reach a majority: in a *minimum* winning block that is no less true of every person on the winning side. The chairperson's franchise does not differ in principle from that of an ordinary member.

It should be said that not all symmetrical games are majoritarian. Example: Consider a 5-person game in which any voting block of exactly two agents has bargaining power 1, and others 0. This game will be symmetrical but is not majoritarian. In the example, of course, there may be more than one winning block. Because winning is associated with bargaining power, and more than one grouping can confer bargaining power, this is not inconsistent.

In practice, voting processes often are based on plurality rule rather than majority rule. For example, elections for the British Parliament, with what is called the "first past the post" standard, and elections in some American states are based on plurality rule. (American states have different election laws and some require a runoff election to determine a majority choice.) For a plurality game we would need a more complex specification, in that the bargaining power (the status as a plurality block) will depend on the partition of the game as a whole. Thus it would require that the voting game be specified as a partition function game. For more detail see McCain (2013, Chapter 3). These complications will be beyond the scope of the chapter.

15.2.2 Linear and TU Games

We will define a linear game as follows: The value to be distributed among the members of the coalition is a constant V. Let the members of the coalition be $C = \{i, j, \ldots, k\}$ and

$$\sum_{i \in C} k_i y_i \leq V \tag{15.1}$$

where y_i is the payout to member i and k_i is an idiosyncratic constant. For a transferable utility, TU, game

$$\sum_{i \in C} y_i \leq V \tag{15.2}$$

That is, $\forall i \in C, k_i = 1$. Clearly a linear game is a generalization of a TU game. In the TU game the coalition simply determines how its surplus is divided among the members of the coalition. We might think of a linear game instead as a mechanism in which the coalition does not determine the payoffs but determines the distribution of a resource, in limited supply, from which the agents produce their payoffs with productivities that vary among them but that are constant for each agent (that is, there are no diminishing returns). The productivity of i is $\frac{1}{k_i}$. Thus, by allocating the resource, the coalition indirectly determines the payoffs of the agents.

For a bargaining power game, the solution is characterized by McCain (2013, Chapter 6)

$$max \sum_{S \subset N} \beta_S \ln(g(S)) \tag{15.3}$$

subject to appropriate constraints, where $g(S)$ is the payoff to group S in excess of the disagreement point for group S. For a voting game, however, either $\beta_S = 1$ if S is a minimal winning voting block or $\beta_S = 0$ otherwise. For this chapter we will assume that for $|S| > 1$, the disagreement point is expressed simply by a nonnegativity constraint. That is, no group of two or more agents can gain any benefit from working together apart from their participation in a minimal willing coalition.

15.2.3   A TU Game with Nonnegativity Constraints

We will proceed by steps from the simplest case, a TU game, that is, a game with $k_i = 1$ for each i. Initially we also suppose that individual rationality constraints are nonnegativity constraints.

We may consider three kinds of solutions both for this and the more complex cases. First, the solution might be *equalitarian*, that is, $y_i = y_j$ for all i, j in the voting population N. Second, the solution might be a *utilitarian* one. For the purposes of this chapter, a utilitarian solution is one that maximizes a weighted sum of the utility payouts $\sum_{i \in N} \lambda_i y_i$. That is, a utilitarian solution is one that is Pareto-optimal, since utilities are subjective and noncomparable. We might define a *classical* solution as one in which the payoffs are proportional to the productivities (where productivity differences among individuals are defined, as in a linear game). Interpreting the linear game as one in which a coalitional mechanism distributes a resource which the individuals transform into payoffs, a classical solution corresponds to equal distribution of the resource. Perhaps we may think of this as an idealization of "equal opportunity," if opportunity is a divisible resource.

The case of a TU game with nonnegativity rationality constraints is characterized in equations (A15.10a)–(A15.11c) in the appendix to this chapter. Equation (A15.7a) (substituting $k_i = 1$) establishes that the bargaining solution is utilitarian. (This is hardly surprising since the bargaining game assures the Pareto-optimality of the result but deserves mention in this specific context.) It follows from equation (A15.10b) that in a TU game, the distributive weights λ_i are equal for all i. This does not quite establish that the distribution will be egalitarian in this case, but this can be established by enumeration, based on the symmetry of the majoritarian voting-bargaining game (see equations A15.11a–c). For the classical solution, since the productivities are equal, the classical and equalitarian solutions coincide. Thus we have a triple coincidence for this simple case: the voting-bargaining solution is at once utilitarian, equalitarian, and classical.

15.2.4   Linear Games in General

For linear games in general, it can be shown that a symmetrical voting solution for such a game will not be equalitarian. For now, continue to assume that the individual rationality constraints are as $v_i = 0$. This case is characterized in the appendix by equations (A15.6a)–(A15.7b). From equations (A15.3a) and (A15.7b) it is established that the voting-bargaining solution is utilitarian. However, it will not be equalitarian, since equations (A15.7a–b) are inconsistent with (A15.11a–b). Further, as equations (A15.8)–(A15.9d) show, neither is it classical.

Now consider a more general case in which there are positive, idiosyncratic individual disagreement points. That is, we suppose that, in case there is no bargain and the game proceeds to a contested election, individual i can expect a payoff v_i. This case is more complex, and the mathematical development comprises equations (A15.1a)–(A15.5d). As this case subsumes the previous one, we can again say that the solution will be neither equalitarian nor classical. Expression (A15.3c) establishes that it is utilitarian.

The idiosyncratic threat points v_i presumably represent the benefits to decision maker i expected in the case of a contested election. As an example, suppose that the proposal is for a tax reform, and the negotiation has to do with the details of the reform. If the bargaining were to fail, so that the proponent group fail to recruit the votes necessary to pass, then the disagreement points v_i would correspond to the status quo. More generally, however, the outcome of a contested election might be uncertain, and v_i interpreted as an expected utility, with risk aversion playing a role in its valuation.

15.2.5 TU Games with Idiosyncratic Rationality Constraints

Now consider the special case of a TU game, in which there are idiosyncratic individual rationality constraints $v_i \geq 0$. As before, of course, the solution is utilitarian. This follows from equations (A15.3b) and (A15.3c). The solution will not be equalitarian in this case, but will approximate an equalitarian solution in a certain sense. This is expressed in Theorem 1 in the appendix to this chapter. The voting members will be divided into two groups, with those in the first group having the lesser and those in the second group having the greater individual rationality constraints. The first group, with the larger rationality constraints, will get those rationality constraints and no more. Once those with the larger rationality constraints have been made whole, the rest is distributed among the other group so that they all have equal payoffs. This could be expressed by saying that the result of the voting-bargaining process, in this case, is to equalize payoffs from the bottom up.

If we think of the negotiation as one for a tax reform or similar measure, so that the disagreement points are determined by the status quo, then we may put it this way: those most advantaged under the status quo neither gain nor lose by the bargaining outcome, while those less advantaged are raised to an equal status equivalent to the least advantaged in the first, more advantaged group.

15.2.6 Visualization and Imperfections

The payout rule set out in Theorem 1 in the appendix to this chapter is visualized in Figure 15.1. In the figure, we assume a coalition of 40 members with the individual rationality constraints (second-best alternatives) shown by the black curve, and the payouts are shown by the broad gray curve, assuming that the surplus created by the coalition is equal to the area of the irregular gray area between them.

It should be stressed that this is an idealized result, in two senses.

First, to implement this rule for payouts would require a great deal of detailed information on the idiosyncratic rationality constraints. This information is likely to be costly and in that case "optimally imperfect" coalitional decisions would at best approximate this rule through easier, heuristic rules. Nevertheless, the model corresponds to the intuition that majoritarian institutions will tend to equalize payoffs mainly by raising the payoffs to the poorest.

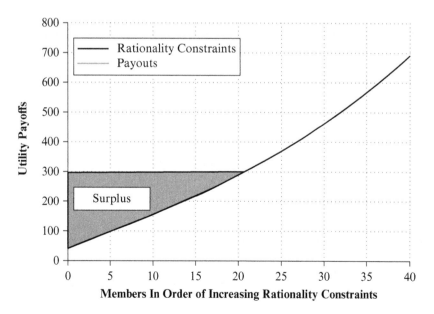

Figure 15.1   Rationality constraints and payouts

Second, and more fundamentally, the argument that majoritarian decisions are symmetrical assumes that any minimal winning voting block can form. Ethnic divisions and political parties that exclude some winning voting blocks will create a more complex situation in which the analysis based on symmetry will not apply. We might think of a political party as an agreement among a group of voters that no one of them will join a voting block unless they all do. If the political party has a majority, then no other winning voting block can exist. This means that only party members obtain any share of the coalitional surplus. Even a minority party can enhance the bargaining power of its members. Consider a five-person example with voters a, b, c, d, e, with the rationality constraints at zero. Suppose ab form a political party. This rules out six of the ten possible minimum winning blocks and results in a situation that a and b share the entire surplus, even though they have to recruit one of the other agents to achieve a majority. Suppose then that d and e respond by forming a political party. Now eight of the ten winning blocks are excluded, but the result is that the "swing vote," agent c, obtains the entire surplus. Since d and e gain nothing by forming a party among themselves, c cannot expect to be a free rider, and may instead be attracted to form a majority party with d and e, and so obtain a third of the surplus. Since unequal payoffs are prevented by the bargaining solution, this majority will be stable – unless, of course, the political parties can commit themselves to more complex agreements. In any case, it appears that the organization of a voting group into political parties will have decided, but unpredictable, influences on their bargaining power.

15.3 "SHAREHOLDER DEMOCRACY"

Voting plays an ostensible role in the governance of for-profit corporations in the United States and elsewhere. In principle, the designation of directors and some other decisions are made by a canvas of the shareholders. Decisions are by weighted voting: that is, the vote cast by each shareholder is weighted by the number of shares she owns. A winning voting block requires that the sum of the voting weights of its members be greater than 0.5.

Notice that a weighted voting mechanism is not a symmetrical mechanism. Let the voting weights be 0.31, 0.2, 0.2, 0.2, 0.09, for a, b, c, d, e respectively. A coalition of a and b is a winning coalition, but a cannot be replaced by any nonmember and neither can be replaced by e. Further, e is not a member of any *minimal* winning voting block: therefore his bargaining power is zero and his share of the surplus will be zero. This example establishes that – in the absence of external regulation or a constitutional rule that cannot be revised within the voting process – profits will *not* be distributed in proportion to ownership.

Since weighted voting is not a symmetrical mechanism, none of the formal results in this chapter apply to it.

15.4 SUMMARY

It seems that decisions in many voting bodies might be described by a two-stage game in which the first stage is a bargaining process and the second is a vote that is often a formality. This does not mean that the voting is irrelevant, but, rather, that it limits the threats that may be made and so influences bargaining power at the first stage. In an idealization of such a two-stage game, majority groups have equal bargaining power, and nonmajority groups have none. This chapter uses a recent extension of bargaining theory that attributes bargaining power to groups as well as individuals and assumes that a minimum winning voting block has bargaining power one and other groups and individuals have bargaining power zero. For TU games, this yields a striking rule for the bargaining solution: the surplus generated by the coalition is either distributed as equal payouts, or distributed among the members with lesser individual rationality constraints, so that their payouts are equal, while others get their individual rationality constraints.

NOTES

1. For a good overview, with some important recent results, see Dasgupta and Maskin, (2008). For a sample of earlier writing with emphasis on cooperative game theory, see Riker (1962). For noncooperative models see Gibbard (1973), Satterthwaite (1975), and Feldman (1979). In earlier discussion, Bowen (1943) should not be neglected. Bowen's paper also gave rise to a very large critical literature.
2. Maskin (1999). This paper was presented in 1977 but not published until twenty years later.
3. This is the simple majority game discussed in von Neumann and Morgenstern (1994/2004, at pp. 222–31).

APPENDIX A15

For the purposes of this chapter, a BP (Bargaining Power) game Q comprises

   I.   an index set $N = \{1, 2, 3, \ldots, n\}$ of agents, $n \geq 3$.

   II.   For each $S \subseteq N$, a constraint function $f_S(\{y_i\}_{i \in S}) \leq 0$, where y_i is the payoff to agent i

       i.   For a transferable utility (TU) game, $f_S(\{y_i\}_{i \in S}) = \sum_{i \in S} y_i - V_S$

       ii.   For a *linear* game, $f_S(\{y_i\}_{i \in S}) = \sum_{i \in S} k_i y_i - V_S$, k_i an idiosyncratic constant and V_S a constant.

   III. For each $S \subseteq N$, an index of *bargaining power* β_S.

   IV. Then Q is a *voting game* if

       i. $\exists \mathscr{S} = \{S \subset N \ni |S| > 1\}$ and $\forall S \in \mathscr{S}, \beta_S = 1$ and $\forall S \notin \mathscr{S}$, $\beta_S = 0$.

The elements of S are winning voting blocks. Winning voting blocks will be assumed to have the following property:

       ii. $S \subset T, S \neq T, S \in \mathscr{S} \Rightarrow T \notin \mathscr{S}$.

       iii. We will define $S_i = \{S \in \mathscr{S} | i \in S\}$.

By iii, if agent i has no voting franchise, $S_i = \varnothing$.

    The game Q is *symmetrical* if, given $i \in N, j \in N, i \neq j, S \in S_i, j \notin S$, then

   V. $T = (S \backslash \{i\}) \cup \{j\} \in \mathscr{S}_j$

That is, a winning block can be sustained by replacing one of its members with any nonmember.

Lemma 1. If Q is symmetrical then for $i \in N, j \in N, i \neq j, S_i| = |S_j|$.

That is, all winning voting blocks are of the same absolute size.

Proof: Let $S_1, S_2, \ldots, S_m$ be an enumeration of S_i. Then if $j \in S_i$, let T_i be S_i. If $j \notin S_i$, let T_i be $(S \backslash \{i\}) \cup \{\varphi\}$. Note that T_i is an element of S_j in either case. Thus, for every element of S_i, there is at least one element of S_j; that is, $|S_j| \geq |S_i|$. Similar reasoning establishes $|S_i| \geq |S_j|$.

Lemma 2. If Q is symmetrical, $S \in S_i$, then $\exists T \in S_j \ni |T| = |S|$.

Proof: trivial.

   VI. Definition: Q is a majority rule game if $S \in \mathscr{S} \Rightarrow |S| > \frac{n}{2}$. Note that, considering IV.ii, either $|S| > \frac{n}{2} + \frac{1}{2}$ or $|S| > \frac{n}{2} + 1$, depending on n, since larger voting blocks are not *minimal* winning blocks.

Lemma 3: Majoritarian Games are symmetrical

Proof: Given $i \in N$, $j \in N$, $i \neq j$, $S \in S_i$, $j \notin S$, let $T = (S\backslash\{i\}) \cup \{j\}$; $|T| = |S|$, so that, since S is a minimal winning block, T also must be.

Lemma 4: There are symmetrical games that are not majoritarian. This is established by an example in the main text of the chapter.

To apply the bargaining power game model to the most general case, a linear game with idiosyncratic individual rationality constraints, we

$$max \sum_{S \subset N} \beta_S \ln(g(S)) \tag{A15.1a}$$

Assume that for $|S| > 1$, the disagreement point is expressed as a nonnegativity constraint. That is, for $S \subset N$, $|S| > 1$, $V_S = 0$. For singletons,

$$y_i \geq v_i \tag{A15.1b}$$

$$g(S) = \sum_{j \in S} y_j - \sum_{j \in S} v_j \tag{A15.1c}$$

Then the threat point for a winning voting bloc S is

$$\sum_{i \in S} y_i \geq \sum_{i \in S} v_i \tag{A15.1d}$$

Thus we

$$max \sum_{S \in \mathcal{S}} \ln \left(\sum_{i \in S} y_i - \sum_{i \in S} v_i \right) \text{ subject to } V_N - \sum_{i \in N} k_i y_i \geq 0$$
$$\text{and } y_i \geq v_i \tag{A15.2a}$$

The Lagrangean function is

$$\mathcal{L} = \sum_{S \in \mathcal{S}} \ln \left(\sum_{j \in S} y_j - \sum_{j \in S} v_j \right) + \mu \left(V_N - \sum_{i \in N} k_i y_i \right)$$
$$+ \sum_{i \in N} \theta_i (y_i - v_i) \tag{A15.2b}$$

A necessary condition for the maximum is

$$\frac{\partial \mathcal{L}}{\partial y_i} = \sum_{S \in \mathcal{S}_i} \frac{1}{\displaystyle\sum_{j \in S} y_j - \sum_{j \in S} v_j} - \mu k_i + \theta_i \leq 0 \tag{A15.2c}$$

We might compare this solution with one that is *utilitarian*, where by a utilitarian solution we mean one that satisfies

$$max \sum_{i \in N} \lambda_i y_i \qquad \text{(A15.3a)}$$

subject to the same constraints, for some set of positive distributive weights λ_i. There are infinitely many such solutions corresponding to different sets of distributional weights. The first-order necessary condition for a utilitarian solution is

$$\frac{\partial \mathcal{L}}{\partial y_i} = \lambda_i - \mu k_i + \theta_i \leq 0 \qquad \text{(A15.3b)}$$

If we set the distributional weights at

$$\lambda_i = \sum_{S \in \mathscr{S}_i} \frac{1}{\sum_{j \in S} y_j - \sum_{j \in S} v_i} \qquad \text{(A15.3c)}$$

and the necessary conditions for a utilitarian solution are satisfied: the voting solution is a utilitarian solution.

Note that if the constraint at (A15.1b) holds as a strict inequality, then $\theta_i = 0$ and the distributional weight λ_i is proportionate to k_i. In cases in which (A15.1b) holds as an equality, we may have $\theta_i > 0$, so that the distributional weight would be increased relative to this proportion to k_i. Suppose that a member of the coalition has no vote in its decisions. Then S_i is a null set, and the summation (15.3c) is an enumeration over a null set and therefore zero. Thus,

$$-\mu k_i + \theta_i \leq 0 \qquad \text{(A15.4a)}$$

If $y_i \geq v_i > 0$, then (A15.4a) must hold as an equality, so

$$\theta_i = \mu k_i > 0 \qquad \text{(A15.4b)}$$

which in turn requires that constraint (A15.1b) holds as an equality. A disenfranchised member of the coalition gets at most her disagreement point.

Now, consider a classical solution for this case. Recall that the "productivity" of i is $\frac{1}{k_i}$. Then suppose, $\forall i$,

$$y_i = \frac{y^*}{k_i} \qquad \text{(A15.5a)}$$

In effect, y^* is the equal division of the "resource" for the classical solution. Then

$$\sum_{i \in N} y_i = V_N = y^* \sum_{i \in N} \frac{1}{k_i} \qquad \text{(A15.5b)}$$

$$y^* = \frac{V_N}{\displaystyle\sum_{i \in N} \frac{1}{k_i}} \tag{A15.5c}$$

However, we cannot exclude the possibility that

$$v_i > \frac{1}{k_i} \frac{V_N}{\displaystyle\sum_{i \in N} \frac{1}{k_i}} \tag{A15.5d}$$

so that for the general case, the classical solution cannot be assured.

Now consider a special case with

$$\forall i \in N, \, v_i = 0 \tag{A15.6a}$$

and assume an interior solution. Then $\forall i$, $\theta_i = 0$ and (A15.3c) becomes

$$\frac{\partial \mathcal{L}}{\partial y_i} = \sum_{S \in \mathcal{S}_i} \frac{1}{\displaystyle\sum_{i \in S} y_i} - \mu k_i \leq 0 \tag{A15.6b}$$

We know from the general case that the solution is utilitarian. The first-order necessary condition for a utilitarian solution is

$$\frac{\partial \mathcal{L}}{\partial y_i} = \lambda_i - \mu k_i \leq 0, \tag{A15.7a}$$

with, again, the equality in case $y_i > 0$. Evidently we may choose the distributional weights

$$\lambda_i = \sum_{S \in \mathcal{S}_i} \frac{1}{\displaystyle\sum_{i \in S} y_i} \tag{A15.7b}$$

and we find that the voting solution is a particular instance of a utilitarian solution, in which each person has bargaining power that depends on the winning voting blocs to which she or he is a necessary member.

However, the solution will not be equalitarian. For if $y_i = y^*$, $\forall i \in N$,

$$k_i = \frac{1}{\mu} \sum_{S \in \mathcal{S}_i} \frac{1}{|S|y^*} = \frac{1}{\mu} \frac{|\mathcal{S}_i|}{|S|y^*} \tag{A15.8}$$

In (A15.8) the RHS is a constant, from Lemmas 1, 2 and expression (A15.6b). Thus, the constancy of k_i is a necessary condition for an equalitarian distribution in this case. An equalitarian distribution will not be observed in general.

Nor will the distribution be classical. From (A15.7a) we have

$$\mu = \frac{\sum\limits_{i \in N} \lambda_i}{\sum\limits_{i \in N} k_i} \qquad \text{(A15.9a)}$$

We may take $\sum_{i \in N} \lambda_i = 1$ as a normalizing assumption without loss of generality. Then

$$\lambda_i = \frac{k_i}{\sum\limits_{j \in N} k_j} = \sum\limits_{S \in \mathcal{S}_i} \frac{1}{\sum\limits_{j \in S} y_j} \qquad \text{(A15.9b)}$$

The classical solution has been characterized in (A15.5a–c) above. Using these

$$\sum\limits_{j \in S} y_j = V_N \frac{\sum\limits_{j \in S} \dfrac{1}{k_j}}{\sum\limits_{j \in N} \dfrac{1}{k_j}} \qquad \text{(A15.9c)}$$

which says (as intuition would suggest) that for a classical distribution, any group receives a proportion of the coalitional value that is the same as their proportion of total productivity.

Using (A15.9b) with (A15.9c),

$$k_i = |\mathcal{S}_i| V_N \frac{\sum\limits_{j \in S} \dfrac{1}{k_j}}{\sum\limits_{j \in N} \dfrac{1}{k_j}} \sum\limits_{j \in N} k_j \qquad \text{(A15.9d)}$$

Again using Lemmas 1, 2, the right-hand side of (A15.9d) is a constant for all $i \in \Sigma \in \Sigma_i$. Furthermore for any $j \notin S$, $\exists \iota \in S$, $\iota \neq i$, $\ni T = S\backslash\{\iota\} \cup \{j\}$; $T \in \mathcal{S}_i$ and so $k_j = k_i$. Thus, k_i is a constant for all N. We conclude that the constancy of k_i is a necessary condition for the distribution to be classical.

Now consider the still more special case of a TU game. A TU game is a special case of a linear game in which $k_i = k$ for all agents, and without loss of generality $k = 1$. Again assuming that the rationality constraints are satisfied by nonnegativity, we may thus immediately write in place of (A15.2c)

$$\frac{\partial \mathcal{L}}{\partial y_i} = \sum\limits_{S \in \mathcal{S}_i} \frac{1}{\sum\limits_{i \in S} y_i} - \mu \leq 0 \qquad \text{(A15.10a)}$$

That is,

$$\sum_{S \in \mathcal{S}_i} \frac{1}{\sum_{i \in S} y_i} = \lambda_i = \mu \tag{A15.10b}$$

The solution of a TU game with zero threat points is a utilitarian solution with all distributional weights equal. This suggests, but does not quite establish, that the solution to a TU game will be egalitarian, that is, $y_i = y_j$, $\forall\ i,j$. If that is so, then $y_i = \frac{V_N}{n}$. We have

$$\sum_{S \in \mathcal{S}_i} \frac{1}{\sum_{i \in S} y_i} = \sum_{S \in \mathcal{S}_i} \frac{1}{\frac{V_N}{n}|S|} \tag{A15.11a}$$

Consider $\iota \neq i$. By symmetry, for $S \in S_i \ni T \in S_\iota \ni |S| = |T|$; thus by enumeration,

$$\sum_{T \in \mathcal{S}_\iota} \frac{1}{\frac{V_N}{n}|T|} = \sum_{S \in \mathcal{S}_i} \frac{1}{\frac{V_N}{n}|S|} \tag{A15.11b}$$

Set

$$\mu = \sum_{T \in \mathcal{S}_\iota} \frac{1}{\frac{V_N}{n}|T|} \tag{A15.11c}$$

and the necessary conditions for the maximum are fulfilled. The convexity of the logarithm function assures us of the sufficient conditions and uniqueness.

Now consider a case of a TU game with positive idiosyncratic individual threat points $v_i > 0$. Consider a partition of $N = L \cup M$. Referring to expression (A15.2c) above, let $L = \{i \in N \ni \theta_i = 0\}$ and $M = \{i \in N \ni \theta_i > 0\}$. Then let

$$V^* = V - \sum_{j \in M} v_i \text{ and } y = \frac{V^*}{|L|} \tag{A15.12a}$$

For $i \in L$, assign $y_i = y$, and for $i \in M$, assign $y_i = v_i$. Then, adapting (A15.7b) and (A15.11a), and noting that $\sum_{j \in S \cap M} y_i - \sum_{j \in S \cap M} v_i = 0$

$$\sum_{S \in \mathcal{S}_i} \frac{1}{\sum_{j \in S} y_j - \sum_{j \in S} v_j} = \sum_{S \in \mathcal{S}_i} \left[\frac{1}{\frac{V^*}{|L|}|S \cap L|} \right] \tag{A15.12b}$$

Suppose $\exists\ i, \iota \ni i \in S \cap L, \iota \in L|S$. For each $S \in \mathcal{S}_i$, $T = (S|\{i\}) \cup \{\iota\}$ is an element of $\mathcal{S}_\iota$, so that, again, by enumeration,

$$\sum_{S \in \mathscr{S}_i} \frac{1}{\sum_{j \in S} y_j - \sum_{j \in S} v_j} = \sum_{S \in \mathscr{S}_\iota} \frac{1}{\sum_{j \in S} y_j - \sum_{j \in S} v_j} = \theta_i = \theta_\iota = \mu \quad \text{(A15.12c)}$$

Thus, again, payments defined in this way will satisfy the necessary conditions for the maximum.

Further, if $y_i = v_i$, then $v_\iota > y$. The proof follows. Suppose that for $\iota \in N$, $y_\iota = v_\iota$. From (A15.2c) and (A15.7b) with $k_i = k_\iota = 1$,

$$\lambda_i - \mu = 0 \quad \text{(A15.13a)}$$

$$\lambda_\iota - \mu + \theta_\iota = 0 \quad \text{(A15.13b)}$$

From the definition of M, $\theta_\iota > 0$, so

$$\lambda_\iota = \mu - \theta_\iota < \mu = \lambda_i \quad \text{(A15.13c)}$$

Define $\mathscr{S}_1 = \mathscr{S}_i \cap \mathscr{S}_\iota$, $\mathscr{S}_2 = \mathscr{S}_i \backslash \mathscr{S}_\iota$ and $\mathscr{S}_3 = \mathscr{S}_\iota \backslash \mathscr{S}_i$. For each $S \in \mathscr{S}_2$, by symmetry $(S \backslash \{i\}) \cup \{\iota\} \in \mathscr{S}_3$, so, again by enumeration, $|\mathscr{S}_3| = |\mathscr{S}_2|$ and

$$\lambda_i = \sum_{S \in \mathscr{S}_2} \frac{1}{\sum_{j \in S} y_i - \sum_{j \in S} v_i} + \sum_{S \in \mathscr{S}_1} \frac{1}{\sum_{j \in S} y_i - \sum_{j \in S} v_i}$$

$$+ \sum_{S \in \mathscr{S}_3} \frac{1}{|\mathscr{S}_3|(y - v_i) + \sum_{j \in S} y_i - \sum_{j \in S} v_i} + \sum_{S \in \mathscr{S}_\iota} \frac{1}{\sum_{j \in S} y_i - \sum_{j \in S} v_i} \quad \text{(A15.13d)}$$

$$\lambda_\iota = \sum_{S \in \mathscr{S}_3} \frac{1}{\sum_{j \in S} y_i - \sum_{j \in S} v_i} + \sum_{S \in \mathscr{S}_1} \frac{1}{\sum_{j \in S} y_i - \sum_{j \in S} v_i} \quad \text{(A15.13e)}$$

if $y \geq v_i$, it follows that $\lambda_\iota \leq \lambda_i$, a contradiction; therefore $y < v_i$.

In summary, the surplus generated by the coalition is $V - \sum_{i \in N} v_i = V - \sum_{i \in L} v_i - \sum_{i \in M} v_i = V^* - \sum_{i \in L} v_i$. In a symmetrical voting solution, this surplus will be distributed in such a way as to make equal payouts to those with lesser individual threat points, raising them to an equal level, while paying the rationality constraints of others. In effect, the majoritarian coalition distributes the surplus in such a way at to raise the bottom of the income distribution in the coalition. It will be helpful to summarize all this as a theorem:

Theorem 1. For a TU game with idiosyncratic individual rationality constraints, a symmetrical voting-bargaining solution will allocate the coalitional surplus to agents with threat points less than a critical value, in such a way as to equalize their payouts, while other agents will receive their rationality constraints. For a TU game with nonnegativity individual rationality constraints, the solution is utilitarian, classical and equalitarian. For linear games, the solution is utilitarian but not in general either classical or equalitarian.

References

Abreu, Dilip, Prajit K. Dutta and Lones Smith (1994), "The Folk Theorem for Repeated Games: A New Condition," *Econometrica*, **64** (4) (July), 939–48.

Andreoni, James and Emily Blanchard (2006), "Testing Subgame Perfection Apart from Fairness in Ultimatum Games," *Experimental Economics*, **9** (4), 307–21.

Anesi, Vincent and Phillippe de Donder (2013), "A Coalitional Theory of Unemployment Insurance and Employment Protection," *Economic Theory*, **52**, 941–77.

Aoki, M. (1980), "A Model of the Firm as a Stockholder–Employee Cooperative Game," *American Economic Review*, **70** (4) (September), 600–610.

Arrow, Kenneth J. (1951), *Social Choice and Individual Values*, New York: Wiley.

Aumann, Robert J. (1973), "Disadvantageous Monopolies," *Journal of Economic Theory*, **6** (1) (February), 1–11.

Aumann, Robert J. (1974), "Subjectivity and Correlation in Randomized Strategies," *Journal of Mathematical Economics*, **1**, 67–96.

Aumann, R.J. (1987), "Correlated Equilibrium as an Expression of Bayesian Rationality," *Econometrica*, **55**, 1–18.

Aumann, R.J. (1997), "Rationality and Bounded Rationality," *Games and Economic Behavior*, **21**, 2–14.

Aumann, Robert J. (2003), "Presidential Address," *Games and Economic Behavior*, **45**, 2–14.

Aumann, R.J. (2004), "Address," Second World Congress of the Game Theory Society, Marseilles.

Aumann, R.J. (2005), "War and Peace (Nobel Prize Lecture)," The Sveriges Riksbank Prize in Economic Sciences in Memory of Alfred Nobel 2005, available at: http://nobelprize.org/nobel_prizes/economics/laureates/2005/aumann-lecture.html, as of June 9, 2007.

Aumann, R.J. and J.H. Dreze (1974), "Cooperative Games with Coalition Structure," *International Journal of Game Theory*, **3**, 217–37.

Aumann, Robert and Michael Maschler (1964), "The Bargaining Set for Cooperative Games," in M. Dresher, L.S. Shapley and A.W. Tucker (eds), *Advances in Game Theory*, Annals of Mathematics Studies Number 52, Princeton: Princeton University Press, pp. 443–76.

Aumann, Robert J. and Michael Maschler (1972), "Some Thoughts on the Minimax Principle," *Management Science*, **18** (5) (January), 54–63.

Aumann, Robert J. and Michael Maschler (1985), "Game Theoretic Analysis of a Bankruptcy Problem from the Talmud," *Journal of Economic Theory*, **36** (2) (August), 195–213.

Axelrod, Robert (1981), "The Emergence of Cooperation among Egoists," *American Political Science Review*, **75** (2) (June), 306–18.

Axelrod, Robert (1984), *The Evolution of Cooperation*, New York: Basic Books.

Azariadis, Costas (1981), "Self-Fulfilling Prophesies," *Journal of Economic Theory*, **25**, 380–96.

Baharad, Eyal and Zvika Neeman (2002), "The Asymptotic Strategy Proofness of Scoring and Condorcet Consistent Rules," *Review of Economic Design*, **7**, 311–40.

Baumol, W. and R. Quandt (1964), "Rules of Thumb and Optimally Imperfect Decisions," *American Economic Review*, **54** (2) (March), 23–46.

Baumol, William J., John C. Panzar and Robert D. Willig (1982), *Contestable Markets and the Theory of Industry Structure*, New York: Harcourt Brace Jovanovich.

Berg, Joyce, John Dickhaut and Kevin McCabe (1995), "Trust, Reciprocity, and Social History," *Games and Economic Behavior*, **10** (1), 122–42.

Bernheim, B. Douglas (1984), "Rationalizable Strategic Behavior," *Econometrica*, **52** (4) (July), 1007–28.

Bernheim, B. Douglas, Bezalel Peleg and Michael D. Whinston (1987), "Coalition-Proof Nash Equilibria I. Concepts," *Journal of Economic Theory*, **42**, 1–12.

Bowen, Howard (1943), "The Interpretation of Voting in the Allocation of Economic Resources," *Quarterly Journal of Economics*, **58**, 27–48.

Brandenburger, Adam and Barry J. Nalebuff (1997), *Co-Opetition: 1. a Revolutionary Mindset that Redefines Competition and Cooperation; 2. The Game Theory Strategy that's Changing the Game of Business*, Doubleday.

Brandenburger, Adam M. and Harborne W. Stuart, Jr. (1996), "Value-Based Business Strategy," *Journal of Economics and Management Strategy*, **5** (1) (Spring), 5–24.

Brandenburger, Adam and Harborne Stuart (2007), "Biform Games," *Management Science*, **53** (4) (April), 537–49.

Bray, Marianne (2003), "Arm-Wrestle Settles Network Battle," CNN. com, available at: http://edition.cnn.com/2003/WORLD/asiapcf/auspac/

03/10/offbeat.nz.wrestle/, as of November 15, 2007, originally posted on 3/11/2003.

Broder, John (2007), "Governors Join in Creating Regional Pacts on Climate Change," *New York Times*, available at: http://www.nytimes.com/2007/11/15/washington/15climate.html?em&ex=1195275600&en=45651e5591715bd9&ei=5087%0A, as of November 15, 2007, originally posted on 11/15/2007.

Brothwell, Don (1987), *The Bog Man and the Archeology of People*, Cambridge, MA: Harvard University Press.

Brown, A.C. (1975), *A Bodyguard of Lies*, New York: Harper and Row.

Camerer, Colin (1987), "Do Biases in Probability Judgement Matter in Markets: Experimental Evidence," *American Economic Review*, **77** (5) (December), 981–97.

Carraro, Carlo (2003), *The Endogenous Formation of Economic Coalitions*, Cheltenham, UK and Northampton, MA, USA: Edward Elgar.

Carter, J.R. and M.D. Irons (1991), "Are Economists Different, and If So, Why?" *Journal of Economic Perspectives*, **5**, 171–7.

Cass, David and Karl Shell (1983), "Do Sunspots Matter?" *Journal of Political Economy*, **91** (2), 193–227.

Chatain, Olivier and Peter Zemsky (2007), "The Horizontal Scope of the Firm: Organizational Tradeoffs vs. Buyer–Supplier Relationships," *Management Science*, **53** (4) (April), 550–65.

Chatterjee, Satyajit, Russell Cooper and B. Ravikumar (1993), "Strategic Complementarity in Business Formation: Aggregate Fluctuations and Sunspot Equilibria," *The Review of Economic Studies*, **60** (4) (October), 795–811.

Chwe, Michael Suk-Young (1994), "Farsighted Coalitional Stability," *Journal of Economic Theory*, **63** (2) (August), 229–325.

Colby, Bonnie G. (2000), "Cap-and-Trade Policy Challenges: A Tale of Three Markets," *Land Economics*, **76** (4) (November), 638–58.

Coleman, Andrew M. (2003), "Cooperation, Psychological Game Theory, and Limitations of Rationality in Social Interaction," *Behavioral and Brain Sciences*, **26** (2) (April), 139–53.

Conlon, John R. (1993), "Can the Government Talk Cheap? Communication, Announcements, and Cheap Talk," *Southern Economic Journal*, **60** (2) (October), 418–29.

Cooper, Russell W., Douglas V. DeJong, Robert Forsythe and Thomas W. Ross (1990), "Selection Criteria in Coordination Games: Some Experimental Results," *American Economic Review*, **80** (1) (March), 218–33.

Cramton, Peter (1997), "Spectrum Auctions: An Early Assessment," *Journal of Economics and Management Strategy*, **6** (3) (Fall), 431–96.

Cross, J.G. (1965), "A Theory of the Bargaining Process," *American Economic Review*, **55** (March), 67–94.

Dasgupta, Partha and Eric Maskin (2008), "On the Robustness of Majority Rule," *Journal of the European Economic Association*, **6** (5), 949–73.

Davis, M. and M. Maschler (1965), "The Kernel of a Cooperative Game," *Naval Research Logistics Quarterly*, **12**, 223–59.

Dawes, Robyn M. (1980), "Social Dilemmas," *Annual Review of Psychology*, **31**, 169–93.

Debreu, G. and Herbert E. Scarf (1963), "A Limit Theorem on the Core of an Economy," *International Economic Review*, **4** (3) (September), 235–46.

Dresher, M., A.W. Tucker and P. Wolfe (1957), *Contributions to the Theory of Games, Volume III*, Annals of Mathematics Studies, Number 39, Princeton: Princeton University Press.

Dresher, M., L. Shapley and A.W. Tucker (1964), *Advances in Game Theory*, Annals of Mathematics Studies, Number 52, Princeton: Princeton University Press.

Dutta, Prajit (1999), *Strategies and Games: Theory and Practice*, Cambridge, MA: MIT Press.

Elster, Jon (1977), "Ulysses and the Sirens: A Theory of Imperfect Rationality," *Social Science Information*, **16** (October), 469–526.

Famous-inventors.com (2006), "Biography of Garrett Morgan," available at: http://www.famous-inventors.com/biography-of-garrett-morgan. html, as of November 15, 2007.

Fehr, E. and Urs Fischbacher (2004), "Third-Party Punishment and Social Norms," *Evolution and Human Behavior*, **25**, 63–87.

Feldman, Allan (1979), "Manipulating Voting Procedures," *Economic Inquiry*, **17** (July), 452–74.

Foley, Duncan K. (1967), "Resource Allocation in the Public Sector," *Yale Economic Essays*, **7** (Spring), 73–6.

Forgo, Ferenc, Jeno Szep and Ferenc Szidarovszky (1999), *Introduction to the Theory of Games: Concepts, Methods, Applications*, Dordrecht: Kluwer.

Foster, Dean P. and Rakesh V. Vohra (1997), "Calibrated Learning and Correlated Equilibrium," *Games and Economic Behavior*, **21** (12) (October–November), 40–55.

Frey, Bruno (2008), *Happiness: A Revolution in Economics*, Cambridge, MA: MIT Press.

Friedman, Daniel (1998), "Evolutionary Economics Goes Mainstream: A Review of the Theory of Learning in Games," *Evolutionary Economics*, **8** (4), 423–32.

Fudenberg, Drew and David K. Levine (1981), "Perfect Equilibria of

Finite and Infinite Horizon Games," UCLA Department of Economics Working Paper # 216.

Fudenberg, Drew and David K. Levine (1999), "Conditional Universal Consistency," *Games and Economic Behavior*, **29**, 104–30.

Fudenberg, D. and E. Maskin (1986), "The Folk Theorem in Repeated Games with Discounting and with Incomplete Information," *Econometrica*, **54**, 533–54.

Galbraith, John Kenneth (1973), *Economics and the Public Purpose*, Houghton Mifflin.

Gale, David and L.S. Shapley (1962), "College Admissions and the Stability of Marriage," *American Mathematical Monthly*, **69**, 9–14.

Gardner, Roy (2003), *Game Theory for Business and Economics*, 2nd Edition, Hoboken: Wiley.

Gibbard, Alan (1973), "Manipulation of Voting Schemes: A General Result," *Econometrica*, **41** (4), 587–601.

Gifford, Daniel and Robert Kudrle (2010), "The Law and Economics of Price Discrimination in Modern Economics: Time for Reconciliation?", *Law Review of the Law School of the University of California at Davis*, **43** (4), 1235–94.

Gillies, D.B. (1953), "Some Theorems on n-Person Games," Doctoral Dissertation, Princeton University, Princeton, NJ.

Gintis, Herbert (2007), "A Framework for the Unification of the Behavioral Sciences," *Behavioral and Brain Sciences*, **30** (1), 1–61.

Glenn, David (2007), "3 Americans Win Nobel Prize in Economics," *The Chronicle of Higher Education*, October 26, p. 16.

Goolsbee, Austan, Steven Levitt and Chad Syverson (2013), *Microeconomics*, New York: Worth.

Greenberg, Joseph (1990), *The Theory of Social Situations: An Alternative Game-Theoretic Approach*, Cambridge University Press.

Greenberg, J. (1994), "Coalition Structures," in R.J. Aumann and S. Hart (eds), *Handbook of Game Theory*, Elsevier, pp. 1305–37.

Groves, T. and J. Ledyard (1977), "Optimal Allocation of Public Goods: A Solution to the 'Free Rider' Problem," *Econometrica*, **45** (4) (May), 783–809.

Guth, W., R. Schmittberger and B. Schwartze (1982), "An Experimental Analysis of Ultimatum Bargaining," *Journal of Economic Behavior and Organization*, **3**, 376–88.

Haag, Matthew and Roger Lagunoff (2007), "On the Size and Structure of Group Cooperation," *Journal of Economic Theory*, **135** (1) (July), 68–89.

Hahn, Robert (1989), "Economic Prescriptions for Environmental Problems: How the Patient Followed the Doctor's Orders," *Journal of Economic Perspectives*, **3** (2) (Spring), 95–114.

Hall, Robert E. (2008), "General Equilibrium with Customer Relationships: A Dynamic Analysis of Rent-Seeking," Hoover Institution, available at: http://www.stanford.edu/~rehall/Recent_Unpublished_Papers.html, as of August 21, 2011.

Hall, Robert E. and Paul R. Milgrom (2008), "The Limited Influence of Unemployment on the Wage Bargain," *American Economic Review*, **98** (4), 1653–74.

Hanson, Niel (2001), *The Custom of the Sea*, John Wiley and Sons.

Hargreaves Heap, Shaun P. and Yanis Varoufakis (1995), *Game Theory: A Critical Introduction*, London: Routledge.

Harris, Robert and Jeremy Paxman (2002), *A Higher Form of Killing: The Secret History of Chemical and Biological Warfare*, New York: Random House Trade Paperbacks.

Harsanyi, John C. (1956), "Approaches to the Bargaining Problem before and after the Theory of Games: A Critical Discussion of Zeuthen's, Hicks', and Nash's Theories," *Econometrica*, **XXIV** (April), 144–57.

Harsanyi, John (1963), "A Simplified Bargaining Model for the n-Person Cooperative Game," *International Economic Review*, **4** (2) (May), 194–220.

Harsanyi, John C. (1967–68), "Games with Incomplete Information Played by 'Bayesian' Players," *Management Science*, **14**, Parts I, 159–83; II, 320–34; III, 486–502.

Harsanyi, John (1975), "Can the Maximin Principle Serve as a Basis for Morality? A Critique of John Rawls' Theory," *American Political Science Review*, **69**, 594–606.

Harsanyi, John and Reinhard Selten (1972), "A Generalized Nash Solution for Two-Person Bargaining Games with Incomplete Information," *Management Science*, **18** (5) (January), 80–106.

Harsanyi, John and Reinhard Selten (1988), *A General Theory of Equilibrium Selection in Games*, Cambridge, MA: MIT Press.

Hart, Sergiu and Andreu Mas-Colell (2000), "A Simple Adaptive Procedure Leading to Correlated Equilibrium," *Econometrica*, **68** (5) (September), 1127–50.

Henrich, Joseph, Robert Boyd, Samuel Bowles, Colin Camerer, Ernst Fehr, Herbert Gintis, Richard McElreath, Michael Alvard, Abigail Barr, Jean Ensminger, Natalie Smith Henrich, Kim Hill, Francisco Gil-White, Michael Gurven, Frank W. Marlowe, John Q. Patton and David Tracer (2005), "'Economic man' in cross-cultural perspective: Behavioral experiments in 15 small-scale societies," *Behavioral and Brain Sciences*, **28** (6) (December), 795–815.

Hodgson, Geoffrey M. (2002), "Darwinism in Economics: From Analogy to Ontology," *Journal of Evolutionary Economics*, **12** (3) (July), 259–81.

Hurwicz, Leonid (1973), "The Design of Mechanisms for Resource Allocation," *The American Economic Review*, **63** (2) (May), 1–30.

Kaldor, Nicholas (1934), "The Equilibrium of the Firm," *Economic Journal*, **44** (173) (March), 60–76.

Keystone Automobile Club (1927), *Eastern Tours*, Keystone Automobile Club.

Klemperer, Paul (2002), "How (Not) To Run Auctions: The European Telecom Auctions," *European Economic Review*, **46** (45) (April), 829–45.

Knight, Frank (1921), *Risk, Uncertainty and Profit*, Boston: Houghton Mifflin.

Kreps, D.M., Paul Milgrom, John Roberts and R. Wilson (1982), "Rational Cooperation in the Finitely Repeated Prisoners' Dilemma," *Journal of Economic Theory*, **27** (2), 245–52.

Kuhn, H.W. (1997), *Classics in Game Theory*, Princeton: Princeton University Press.

Kuhn, H.W. and A.W. Tucker (1950), *Contributions to the Theory of Games, Volume I*, Annals of Mathematics Studies, Number 24, Princeton: Princeton University Press.

Kuhn, H.W. and A.W. Tucker (1953), *Contributions to the Theory of Games, Volume II*, Annals of Mathematics Studies, Number 28, Princeton: Princeton University Press.

Lamar-Sterling, Sara (2006), "Drawing Straws," Park Avenue Methodist Church, New York, available at: http://www.parkavemethodist.org/sermon.php?s=1, as of November 15, 2007, originally posted on 5/28/2006.

Lave, L.B. (1965), "Factors Affecting Cooperation in the Prisoner's Dilemma," *Behavioral Science*, **10**, 26–38.

Lerner, Abba P. (1934), "The Concept of Monopoly and the Measurement of Monopoly Power," *The Review of Economic Studies*, **1** (3) (June), 157–75.

Lewis, W.A. (1942), "Notes on the Economics of Loyalty," *Economica*, **9** (36) (November), 333–48.

Lindbeck, Assar and Dennis Snower (2001), "Insiders versus Outsiders," *Journal of Economic Perspectives*, **15** (1) (Winter), 165–88.

Lohr, Steve (2007), "Three Share Nobel in Economics for Work on Social Mechanisms," *New York Times*, available at: http://www.nytimes.com/2007/10/16/business/16nobel.html?_r=1&ref=business&oref=slogin, as of October 16, 2007, originally posted on 10/16/2007.

Lucas, W.F. (1968), "A Game With No Solution," *Bulletin of the American Mathematical Society*, **74**, 237–9.

Luce, R. Duncan and Howard Raiffa (1957), *Games and Decisions*, New York: Wiley and Sons.

MacLeod, W. Bentley and James Malcomson (1989), "Implicit Contracts, Incentive Compatibility, and Involuntary Unemployment," *Econometrica*, **57** (2) (March), 447–80.

Mailath, G.J. (1998), "Do People Play Nash Equilibrium? Lessons from Evolutionary Game Theory," *Journal of Economic Literature*, **36** (3), 1347–74.

Market Design, Inc. (2007), "MDI Projects," available at: http://www. market-design.com/projects-telecommunications.html, as of Jan. 6, 2015.

Marwell, Gerald and Ruth E. Ames (1981), "Economists Free Ride: Does Anybody Else?" *Journal of Public Economics*, **15**, 295–310.

Marx, Karl (1845), "Theses on Feuerbach," Marxists.org, available at: http://www.marxists.org/archive/marx/works/1845/theses/index.htm, as of October 20, 2007.

Marx, Karl and F. Engels (1848), "Manifesto of the Communist Party," available at: http://www.marxists.org/archive/marx/works/1848/communist-manifesto/, as of September 7, 2004.

Maskin, E. (1999), "Nash Equilibrium and Welfare Optimality," *Review of Economic Studies* (Paper presented at the summer workshop of the Econometric Society in Paris, June 1977), **66**, 23–38.

Maskin, Eric (2004), "Bargaining, Coalitions and Externalities," Plenary Lecture, Second World Congress of the Game Theory Society, Marseille.

Maskin, Eric and Jean Tirole (1987), "Correlated Equilibria and Sunspots," *Journal of Economic Theory*, **43**, 364–73.

McCain, Roger A. (1972), "Distributional Equality and Aggregate Utility: Further Comment," *American Economic Review*, **62** (3) (June), 497–500.

McCain, Roger A. (1978), "Endogenous Bias in Technical Progress and Environmental Policy," *American Economic Review*, **68** (4) (September), 538–46.

McCain, Roger A. (1980), "A Theory of Codetermination," *Zeitschrift für Nationalökonomie*, **40** (12), 65–90.

McCain, Roger A. (1985), "Economic Planning for Market Economies: The Optimality of Planning in an Economy with Uncertainty and Asymmetrical Information," *Economic Modelling*, **2** (4) (October), 317–23.

McCain, Roger A. (1991), "A Theory of Economic Planning for Market Economies: The Optimality of Planning," in S. Baghwan Dahiya (ed.), *Theoretical Foundations of Development Planning*, Vol. 3, New Delhi: Concept Publishing.

McCain, Roger (2007), "Welfare Economics," in William Darity (ed.),

International Encyclopedia of the Social Sciences, 2nd Edition, Farmington Hills, MI: Gale.

McCain, Roger A. (2013), *Value Solutions in Cooperative Games*, Singapore: World Scientific.

McCain, Roger A. (2014a), *Reframing Economics: Economic Action as Imperfect Cooperation*, Cheltenham, UK and Northampton, MA, USA: Edward Elgar.

McCain, Roger A. (2014b), *Game Theory: A Nontechnical Introduction to the Analysis of Strategy*, 3rd Edition, Singapore: World Scientific.

McCain, Roger A., Richard Hamilton and Frank Linnehan (2011), "The Problem of Emergency Department Overcrowding: Agent-Based Simulation and Test by Questionnaire," in Sjoukje Osinga, Gert Jan Hofstede and Tim Verwaart (eds), *Notes in Economics and Mathematical Systems: Emergent Results of Artificial Economics*, Heidelberg: Springer Verlag, pp. 91–102.

McKelvey, R.D. and T.R. Palfrey (1992), "An Experimental Study of the Centipede Game," *Econometrica*, **60** (4), 803–36.

McKinsey, J.C.C. (1952), *Introduction to the Theory of Games*, New York: McGraw-Hill.

McMillan, John, Michael Rothschild and Robert Wilson (1997), "Introduction (to Special Issue on Market Design and Spectrum Auctions)," *Journal of Economics and Management Strategy*, **6** (3) (Fall), 425–30.

Mill, John Stuart (1898/1965), *Principles of Political Economy*, People's Edition, London: Longmans, Green and co.

Montet, Christian and Daniel Serra (2003), *Game Theory and Economics*, Basingstoke: Palgrave Macmillan.

Morehouse, L.G. (1967), "One-Play, Two-Play, Five-Play and Ten-Play Runs of Prisoner's Dilemma," *Journal of Conflict Resolution*, **11**, 354–62.

Morgenstern, Oskar and G. Schwödiauer (1976), "Competition and Collusion in Bilateral Markets," *Zeitschrift für Nationalökonomie*, **36** (4), 217–45.

Mortensen, Dale T. and Christopher Pissarides (1994), "Job Creation and Job Destruction in the Theory of Unemployment," *Review of Economic Studies*, **61** (3) pp. 397–415.

Moulin, Herve (1982), *Game Theory for the Social Sciences*, New York: NYU Press.

Moulin, H. and B. Peleg (1982), "Cores of Effectivity Functions and Implementation Theory," *Journal of Mathematical Economics*, **10** (1) (June), 115–45.

Myerson, Roger B. (1979), "Incentive Compatibility and the Bargaining Problem," *Econometrica*, **47**, 61–73.

Myerson, R. (1986), "Multistage Games with Communication," *Econometrica*, **54**, 323–58.

Nash, John (1950a), "Equilibrium Points in n-Person Games," *Proceedings of the National Academy of Sciences*, **36**, 48–9.

Nash, John (1950b), "The Bargaining Problem," *Econometrica*, **18**, 155–62.

Nash, John (1951), "NonCooperative Games," *Annals of Mathematics*, **2** (September), 286–95.

Nash, John (1953), "Two-Person Cooperative Games," *Econometrica*, **21** (January), 128–40.

Newell, A. and H.A. Simon (1972), *Human Problem Solving*, Englewood Cliffs, NJ: Prentice-Hall.

New York Times (2014), "Spite Is Good. Spite Works," available at: http://www.nytimes.com/2014/04/01/science/spite-is-good-spite-works.html?_r=0, as of September 25, 2014.

Nutter, Warren (1964), "Duopoly, Oligopoly and Emerging Competition," *Southern Economic Journal*, **30** (April), 342–52.

Oosterbeek, Hessel, Randolph Sloof and Gijs van de Kuilen (2004), "Cultural Differences in Ultimatum Game Experiments: Evidence from a Meta-analysis," *Experimental Economics*, **7** (2) (June), 171–88.

Osborne, Martin (2004), *An Introduction to Game Theory*, Oxford: Oxford University Press.

Pareto, Vilfredo (1906/1971), *Manual of Political Economy*, first published in Italian in 1906, English translation Augustus Kelley 1971.

Pearce, D.G. (1984), "Rationalizable Strategic Behavior and the Problem of Perfection," *Econometrica*, **52** (4) (July), 1029–50.

Peck, James and Karl Shell (1991), "Market Uncertainty: Correlated and Sunspot Equilibria in Imperfectly Competitive Economies," *The Review of Economic Studies*, **58** (5) (October), 1011–29.

Peleg, B. and P. Sudhölter (2003), *Introduction to the Theory of Cooperative Games*, Dordrecht: Kluwer.

Pen, Jan (1952), "A General Theory of Bargaining," *American Economic Review*, **42**, 24–42.

Pigou, A.C. (1920), *Economics of Welfare*, London: Macmillan.

Plott, Charles R. (1997), "Laboratory Experimental Testbeds: Application to the PCS Auction," *Journal of Economics and Management Strategy*, **6** (3) (Fall), 605–38.

Poundstone, William (1992), *Prisoner's Dilemma*, New York: Doubleday.

Rapoport, Anatole and Albert M. Chammah (1965), *Prisoner's Dilemma*, University of Michigan Press.

Rawls, John (1971), *A Theory of Justice*, Cambridge, MA: Belknap Press.

Ray, Debraj and Rajiv Vohra (1999), "A Theory of Endogenous Coalition Structures," *Games and Economic Behavior*, **26**, 286–336.

Riker, William H. (1962), *The Theory of Political Coalitions*, New Haven: Yale University Press.

Robinson, Joan (1933), *The Economics of Imperfect Competition*, London: Macmillan.

Robinson, Joan and John Eatwell (1974), *An Introduction to Modern Economics*, McGraw-Hill.

Rosenstein-Rodan, Paul (1943), "Problems of Industrialization of Eastern and South-Eastern Europe," *Economic Journal*, **53** (June–September), 202–11.

Rosenthal, R. (1981), "Games of Perfect Information, Predatory Pricing, and the Chain Store Paradox," *Journal of Economic Theory*, **25**, 92–100.

Roth, Alvin E. (1979), *Axiomatic Models of Bargaining*, Berlin: Springer-Verlag.

Roth, Alvin E. (1984), "The Evolution of the Labor Market for Medical Interns and Residents: A Case Study in Game Theory," *The Journal of Political Economy*, **92** (6) (December), 991–1016.

Roth, Alvin and Ido Erev (1995), "Learning in Extensive-Form Games: Experimental Data and Simple Dynamic Models in the Intermediate Term," *Games and Economic Behavior*, **8**, 164–212.

Roth, Alvin E., Vesna Prasnikar, Masahiro Okuno-Fujiwara and Shmuel Zamir (1991), "Bargaining and Market Behavior in Jerusalem, Ljubliana, Pittsburgh, and Tokyo: An Experimental Study," *American Economic Review*, **91** (5) (December), 1068–95.

Rousseau, Jean-Jacques (1754), "Discourse on Inequality (G.D.H. Cole Translation), Part II," The Constitution Society, available at: http://www.constitution.org/jjr/ineq_04.htm, as of September 9, 2006.

Royal Swedish Academy of Sciences, "The Official Web Site of the Nobel Prizes," available at: http://www.nobelprize.org/nobel_prizes/economic-sciences/as of January 7, 2015.

Rubinstein, Ariel (1982), "Perfect Equilibrium in a Bargaining Model," *Econometrica*, **50** (1) (January), 97–109.

Samuelson, Paul (1954), "The Pure Theory of Public Expenditures," *Review of Economics and Statistics*, **36** (4) (November), 387–9.

Satterthwaite, M. (1975), "Strategy-Proofness and Arrow's Conditions: Existence and Correspondence Theorems for Voting Procedures and Social Welfare Functions," *Journal of Economic Theory*, **10** (2) (April), 189–203.

Scarf, Herbert E. (1967), "The Core of an N Person Game," *Econometrica*, **35** (1) (January), 50–69.

Schelling, Thomas (1960), *The Strategy of Conflict*, Cambridge, MA: Harvard University Press.

Schelling, T.C. (1978), *Micromotives and Macrobehavior*, New York: Norton.

Schelling, T.C. (1980), "The Intimate Contest for Self-Command," *The Public Interest*, **60** (Summer), 94–118.

Schmeidler, David (1969), "The Nucleolus of a Characteristic Function Game," *SIAM Journal on Applied Mathematics*, **17** (6) (November), 1163–70.

Schumpeter, J.A. (tr. by Redvers Opie) (1934/1911), *The Theory of Economic Development*, Cambridge, MA: Harvard University Press.

Schumpeter, J.A. (1947), "The Creative Response in Economic History," *The Journal of Economic History*, **7** (2) (November), 149–59.

Scitovsky, Tibor de (1941), "A Note on Welfare Propositions in Economics," *Review of Economic Studies*, **9**, 77–88.

Scitovsky, T. de (1943), "A Note on Profit Maximisation and its Implications," *The Review of Economic Studies*, **II** (1), 57–60.

Scitovsky, Tibor (1954), "Two Concepts of External Economies," *The Journal of Political Economy*, **62** (2) (April), 143–51.

Selten, Reinhard (1964), "Valuation of N-Person Games," in M. Dresher, L.S. Shapley and A.W. Tucker (eds), *Advances in Game Theory*, Annals of Mathematics Studies, Number 52, Princeton: Princeton University Press, pp. 577–626.

Selten, Reinhard (1975), "Reexamination of the Perfectness Concept for Equilibrium Points in Extensive Games," *International Journal of Game Theory*, **4**, 25–55.

Selten, Reinhard and G. Gigerenzer (2001), *Bounded Rationality: The Adaptive Toolbox*, Cambridge, MA: MIT Press. Retrieved November 28, 2007 from Drexel University, Hagerty Library, Net Library: http://www.netlibrary.com.

Sen, Amartya (1969), "Quasi-Transitivity, Rational Choice and Collective Decisions," *The Review of Economic Studies*, **36** (3), 381–93.

Sen, Amartya (1970), "The Impossibility of a Paretian Liberal," *Journal of Political Economy*, **78** (1) (January), 152–7.

Serrano, Roberto (2003), "Fifty Years of the Nash Program, 1953–2003," Working Paper, Brown University.

Shapley, L.S. (1953), "Stochastic Games," *Proceedings, National Academy of Sciences*, **39**, 1095–100.

Shapley, Lloyd and Martin Shubik (1952), "Solutions of N-Person Games with Ordinal Utilities (abstract)," *Econometrica*, **21**, 348–9.

Shapley, Lloyd and Martin Shubik (1954), "A Method for Evaluating the Distribution of Power in a Committee System," *The American Political Science Review*, **48** (3) (September), 787–92.

Shapley, Lloyd and Martin Shubik (1969), "On the Core of an Economic

System with Externalities," *American Economic Review*, **59** (4) (September), 678–84.

Shimer, Robert (2005), "The Cyclical Behavior of Equilibrium Unemployment and Vacancies," *American Economic Review*, **95** (1) 25–49.

Shubik, M. (1962), "Incentives, Decentralized Control, the Assignment of Joint Costs and Internal Pricing," *Management Science*, **8** (3) (April), 325–43.

Simon, H.A. (1978), "Rationality as a Process and as a Product of Thought," *American Economic Review*, **68** (2) (May), 1–16.

Simon, H.A. (1995), "Artificial Intelligence: An Empirical Science," *Artificial Intelligence*, **77**, 95–127.

Smith, Adam (1994/1776), *The Wealth of Nations*, New York: The Modern Library.

Smith, John Maynard (1972), "Game Theory and the Evolution of Fighting," in John Maynard Smith (ed.), *On Evolution*, Edinburgh: Edinburgh University Press, pp. 8–28.

Stahl, Dale O. and Paul W. Wilson (1995), "On Players' Models of Other Players: Theory and Experimental Evidence," *Games and Economic Behavior*, **10**, 218–54.

Stanley, T.D. and Ume Tran (1998), "Economics Students Need Not Be Greedy: Fairness and the Ultimatum Game," *Journal of Socio-Economics*, **27** (6), 657–63.

Stein, Jeremy (1989), "Cheap Talk and the Fed: A Theory of Imprecise Policy Announcements," *American Economic Review*, **79**, 32–42.

Strotz, Robert Henry (1956), "Myopia and Inconsistency in Dynamic Utility Maximization," *Review of Economic Studies*, **23** (3), 163–80.

Stuart, H.W. Jr. (2001), "Cooperative Games and Business Strategy," in K. Chatterjee and William Samuelson (eds), *Game Theory and Business Applications*, Norwell, MA: Kluwer, pp. 189–211.

Svejnar, Jan (1986), "Bargaining Power, Fear of Disagreement, and Wage Settlements: Theory and Evidence from U.S. Industry," *Econometrica*, **54** (5) (September), 1055–78.

Telser, Lester (1978), *Economic Theory and the Core*, Chicago: University of Chicago Press.

Telser, Lester (1997), *Joint Ventures of Labor and Capital*, Ann Arbor: University of Michigan Press.

Thrall, R.M. and W.F. Lucas (1963), "N-Person Games in Partition Function Form," *Naval Research in Logistics Quarterly*, **10**, 281–98.

Tsebelis, George (1990), *Nested Games: Rational Choice in Comparative Politics*, Berkeley, CA: University of California Press.

Tucker, A.W. and R.D. Luce (eds) (1959), *Contributions to the Theory of Games, Volume IV*, Annals of Mathematics Studies Number 40, Princeton: Princeton University Press.

Van Huyck, J., R. Battalio and R. Beil (1990), "Tacit Coordination Games, Strategic Uncertainty, and Coordination Failure," *American Economic Review*, **80**, 234–48.

Vickrey, W. (1961), "Counterspeculation, Auctions, and Competitive Sealed Tenders," *Journal of Finance*, **16** (March), 8–37.

von Neumann, John (1928/1959), "On the Theory of Games of Strategy," trans. Sonya Bargmann, in A.W. Tucker and R.D. Luce (eds), *Contributions to the Theory of Games, Volume IV*, Annals of Mathematics Studies Number 40, Princeton: Princeton University Press, pp. 13–42.

von Neumann, John and Oskar Morgenstern (1944/2004), *Theory of Games and Economic Behavior*, 60th anniversary edition, Princeton: Princeton University Press.

Walker, Paul (2012), "Chronology of Game Theory," University of Canterbury, available at: http://www.econ.canterbury.ac.nz/personal_pages/paul_walker/gt/hist.htm as of January 7, 2015.

Waud, Roger (1970), "Public Interpretation of Federal Reserve Discount Rate Changes: Evidence on the 'Announcement Effect'," *Econometrica*, **38**, 231–50.

Weintraub, E. Roy (1992), "Introduction," in E. Roy Weintraub (ed.), *Toward a History of Game Theory*, Durham: Duke University Press, pp. 3–12.

Zermelo, E. (1913), "Uber eine Anwendung der Mengenlehre auf die theorie des schachspiels," *Proceedings, Fifth International Congress of Mathematicians*, **2**, 501–4.

Zeuthen, Frederick (1930), *Problems of Monopoly and Economic Warfare*, London: Routledge and Kegan Paul.

Zhao, J. (1992), "The Hybrid Solutions of an N-Person Game," *Games and Economic Behavior*, **6** (3), 145–60.

Index